ANTIQUE BOTTI

By the same author

A Treasure Hunter's Guide
Bottle Collecting
International Bottle Collectors' Guide
Non-Dating Price Guide to Bottles,
    Pipes and Dolls' Heads
Pebble Polishing
Rock and Gem Polishing
Treasure Hunting for All

Antique Bottles in Colour

EDWARD FLETCHER

*Photography Jeremy McCabe,
Nico Mavroyenis, Frank Welch*

*BLANDFORD PRESS
Poole · Dorset*

First published in 1976
Reprinted 1978

Copyright © 1976 Blandford Press Ltd
Link House, West Street, Poole,
Dorset BH15 1LL

ISBN 0 7137 0793 3

All rights reserved. No part of this book may be reproduced or transmitted in any form or by any means, electronic or mechanical, including photocopying, recording or by any information storage and retrieval system, without permission in writing from the Publisher.

Typeset in 11 on 12pt Bembo by
Woolaston Parker Ltd, Leicester, England
Printed and bound in Great Britain by Cox & Wyman Ltd
London, Fakenham and Reading

Contents

	Introduction	7
1	Early History of Bottlemaking	9
2	Medicine and Pharmaceutical Bottles	33
3	Mineral Water Bottles	43
4	Household Bottles and Jars	49
5	Ink Bottles	52
6	Figural Bottles	54
7	Coloured Glass in Bottlemaking	56
8	Stoneware Bottles	59
9	The Men who made the Bottles	62
	The Colour Plates	65
	Key to Colour Plates	131
	Forming a Collection of Antique Bottles	164
	Book and Magazine List	166

Introduction

The excavation of long-abandoned refuse dumps to recover objects which our nineteenth century forebears discarded as worthless junk has in recent years achieved remarkable popularity in most English-speaking countries throughout the world. Enthusiasm for this unusual activity has been generated by a surge of popular interest in industrial archaeology, nurtured by massive increases in the prices of conventional antiques which are now beyond the pockets of the majority of would-be collectors, and sustained moreover by a wave of nostalgia for the previous century which has swept around the world as the twentieth century stumbles towards its uncertain conclusion. After more than a decade of digging by a weekend workforce approaching two millions in number the array of recovered relics can only be described as staggering. It includes transfer-printed Prattware pot lids, Parisian dolls' heads, fairings, Nailsea glass, public-house mirrors, matchstrikers, clay tobacco pipes, flower-patterned chamber pots, garden ornaments, buttons, ceramic doorknobs, and examples of china tableware and other articles from almost every known nineteenth century pottery.

But it is empty bottles which comprise the largest group of excavated finds, and it is bottles (and other containers) which the majority of enthusiasts collect and display in their homes as the rewards of their labours 'at the diggings'. Wine bottles, beer bottles, spirits bottles, ink bottles, scent bottles, medicine bottles, food bottles – all thrown when empty into the rubbish dumps of nineteenth century cities and towns. Those which survived this rough treatment and were not buried irretrievably beneath twentieth century office blocks, highways, and factories are now being diligently recovered, carefully cleaned, and proudly displayed as genuine antiques; as delightful examples of hand-worked craftmanship; and as historical relics of the beginnings of a vast packaging industry which now provides us with aerosol deodorants, canned beer, and plastic squeegy bottles. Like our nineteenth century forebears we in turn are now relegating our throwaway packages to municipal dumps on the outskirts of our vast conurbations. It remains to be seen whether twenty-first century collectors will forage for them with the same enthusiasm present-day collectors now have for great-grandma's throwaways.

This book is an attempt to show with the aid of colour photography a representative selection of the glass and stoneware bottles, pots, and jars recovered during the ten years of popularity the hobby has so far enjoyed. Most of the examples illustrated have come from dumps in Britain, the United States, and those countries which were part of the British Empire in the reign of Queen Victoria because it is in these countries that almost all of the dump excavations have so far taken place. Proof that similar finds await those who seek them in other countries is provided by the inclusion of a number of photographs showing attractive and collectable bottles from France, Germany, Holland, Italy, and elsewhere. I hope their inclusion here will encourage readers in those countries to find and explore their nineteenth century refuse dumps to recover specimens and promote the growth of this truly fascinating hobby. Equally I hope that experienced collectors in Britain, the United States, Australia, New Zealand, Canada, and South Africa will be spurred to greater efforts at recovery by the illustrations showing bottles from their countries. Many more specimens of comparable beauty and colour are still to be found – perhaps within a few miles of where you are reading this book. Good hunting!

> Edward Fletcher,
> Redcar, Cleveland, England
> January, 1976

I

Early history of bottlemaking

The earliest vessels for water and other liquids were gourds, hollow stones, and tree trunks; later goatskins made waterproof by coating them inside with oil, resin, or bitumen were used for storing wine. However, it was probably the ancient Egyptians who produced the first manufactured containers. These included small earthenware pots for cosmetics and medicines, and large earthenware amphorae for wine. Archaeologists have noted that the same substances used to render goatskins waterproof were often placed in the bottoms of amphorae before they were filled with wine, probably because wine drinkers had become accustomed to the 'flavour' taken on by wine tainted with oil, resin, and bitumen when stored in skins. Egyptian amphorae did not have corks. Some were closed with lids held in place by clay, pitch, or mortar into which a seal was often stamped; others were left open, air being excluded by pouring oil or honey into the neck after filling with wine.

It was not until the Roman era that amphorae were stoppered by loose-fitting corks held in place by pitch. Cork, the bark of a species of evergreen oak which grows abundantly in southern Europe, was also used by the Romans to make military helmets and fishing floats. In addition to introducing the world to a stoppering material which has been used ever since, the Romans were the first winemakers to mark containers with vintage years. They either stamped their amphorae with the name of the Consul in whose year they were filled or they tied dated labels around the necks.

Both Egyptians and Romans practised the art of glassmaking and made a wide range of glass decanters and serving vessels for oil, perfume, and wine. Many of these were core-moulded by repeatedly dipping a core of sand and clay into molten glass and chipping out the mould when the glass had hardened. A few were simple freeblown globes with crude necks, but archaeologists have also discovered several clay moulds of Roman date for square and corrugated decanters, and even one for a figural vessel in the shape of a bunch of grapes. No doubt these glassmaking skills were carried to other parts of Europe as the Romans thrust their frontiers outwards. Certainly they introduced vine growing and the serving of wine in glass decanters to the barbaric Britons soon after 43 A.D., while the sites of many Roman glassworks and potteries have been found throughout those parts of Europe which once formed the Roman Empire.

In the Dark Ages which followed the fall of Rome glassmaking became a lost art in western Europe, known only to a handful of monasteries where the making of window glass was carried on. The secrets of vineyard cultivation and wine making were also kept alive by monks. Writing in England in 731 A.D. Bede recorded that they were 'vineyards in several places in Britain, generally connected with monasteries'. But when monastic institutions were abolished during the reign of Henry VIII vineyards died out in England because there were no monks to cultivate them.

Wine had then to be imported from Continental Europe and some of it was shipped in stoneware bottles made in the Rhine region of what is now Germany. These Rhenish-ware bottles were squat, pot-bellied, and stoppered with loose-fitting corks. Most of them were made with a single handle. They were glazed by throwing several pounds of common salt into the kiln during firing. The salt reacted with minerals in the clay to produce a glassy, mottled surface on the wares. Some of the bottles were adorned before firing with incised designs including crests, dates, monograms, and grotesque masks thought to represent Cardinal Bellarmine, a controversial Italian clergyman who bitterly opposed the Dutch Reform Church. It is by this name – Bellarmine jugs – that these Rhenish-ware flasks are known today. They are thought to have first been manufactured in England, probably by immigrant workers from the Rhine region, in

1. A scene in a medieval European glassworks.

about 1684; but a patent granted in London in 1622 to Thomas Rous and Abraham Cullyn 'for the making of stone potts, stone juggs, and stone botells for the term of fourteen years' might indicate that English potters were making similar wares before the arrival of these immigrants. However, it is more likely that these English 'stone botells' were white pottery Lambeth jugs; a number of these bearing dates from 1637–72 have survived.

A limited amount of glass was made in England during the medieval period, but it was European immigrants who re-introduced glassmaking on a commercial scale in the mid-1500's. They settled in the south of England and were soon producing window glass, tablewares of all descriptions, and some of the first commercial containers. These were tiny, narrow-necked vials used by alchemists and apothecaries for their elixirs and medicines. They were usually free-blown and light green or aqua in colour.

The usual method of making the glass and producing the bottle was as follows: sand, potash, and lime were mixed together in a crucible of clay and fired in a small rectangular furnace. The potash was obtained by burning wood or bracken, and the wood-fired furnaces were built on steep sloping ground to catch as much draught as possible. Broken pieces of glass, known as cullet, were also added to the crucible in order to speed the melting process. When the materials had combined chemically to form glass the molten liquid was cooled slightly until it possessed the consistency of treacle. At this point the glassblower would dip his iron blowpipe into the crucible and take up a small quantity of glass. By blowing down the pipe he would produce a sphere which he then shaped by rolling it on a flat stone or by spinning the rod between his fingers. When he had produced the required shape the bottle was allowed to cool slightly until it was sufficiently hard to crack or shear from the end of the blowpipe. The completed bottle was then placed on a shelf in the flue of the furnace to cool slowly so that the glass annealed and was thus toughened (Fig. 1).

At this time the imported beverages enjoyed by Englishmen were Rhenish wine, Spanish sack, and French claret. Rhenish wine came in the stoneware jugs mentioned above, but sack and claret were

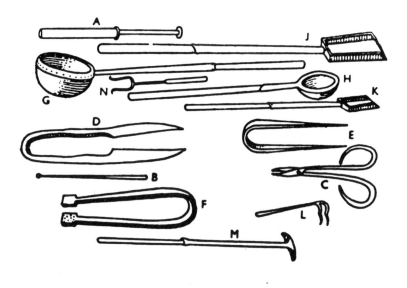

2. Seventeenth century bottlemaking tools. *A* is a blowpipe; *D* and *C* are neck-forming tools.

shipped to England in barrels and allowed to mature. Vintners then drew off the wine into glass bottles which were corked and tied down with pack thread anchored beneath a string-ring encircling the neck an inch or so below the mouth, which was finally dipped into a mixture of resin and pitch to prevent access of air or leakage. (Later red Spanish wax was used.) The bottles were then placed slopewise on sand or sawdust strewn to a thickness of three inches on the floor of a cold cellar. The bottles, of onion shape with long and slender necks, were imported empty from France because until 1592 the secret of making the tougher dark green glass from which wine bottles were blown was unknown in England. In that year glasshouses specialising in the production of these onion-shaped wine bottles were established in London.

The making of wine bottles required a more professional finish to necks and lips than was required for earlier alchemists' and apothecaries' bottles. New tools had to be developed. These included a pontil rod, or punty, to which the bottle could be attached at its base while the neck was worked, and various pliers, shears, and calipers to form the lip (Fig. 2). The method used was as follows. When the glassblower had produced the sphere of glass on the end of his blowpipe an apprentice would take a solid iron pontil rod, or punty, dip it into the molten glass in the crucible, and attach it to the bottom of the sphere. The glassblower then 'wetted off' his blowpipe by letting a few drops of water fall on the neck of the bottle which immediately 'sheared' at that point. Holding the partly-made bottle on the pontil rod he thrust the sheared neck into the furnace until it softened, and then worked it with pliers and calipers before rolling it on a stone or flat piece of iron to produce a neat lip. A ring of glass – the string-ring – was then added below the mouth to facilitate tying of the cork.

These rolled-lip wine bottles were an improvement on earlier sheared-lip medicines but it was difficult to place them squarely on a table or shelf because of the jagged scar left when the pontil rod was broken off. They were therefore wrapped in willow basketwork or placed on a metal ring which lifted the pontil scar clear of the surface and also provided protection for the thin glass. Later, as thicker glass came into use, the 'kick-up' was evolved. This was produced by applying pressure to the pontil rod when attaching it to the bottom of the sphere, or by using the special tool seen in Fig. 4. This produced an indentation, or kick-up.

The annual production figures for these first English bottleworks are not recorded, but it is known that in 1615 James I prohibited the importation of French bottles and that in 1623 he granted monopoly control of 'thirty glasshouses' in various parts of Britain to Sir Robert Mansell. It is also known that in 1636 the sale of wine 'by-the-bottle' was prohibited, probably following complaints from potters about loss of trade to glass bottle makers and from other industries which resented the consumption of vast amounts of timber as fuel for furnaces. This ban on the sale of bottled wine was a severe blow to the embryonic bottlemaking industry. A resulting shortage of bottles led to the practice, by those who could afford it, of owning one's own

Scenes in a French glasshouse in 1750.
3. Blowing a sphere of glass in a cup-mould to achieve cylindrical sides on the bottle.
(continued)

bottles which were used both to carry wine from the vintner's premises and to decant drinks during meal-time. These bottles-cum-decanters were regarded as status symbols and many purchasers adopted the habit of having marks of identification placed on them during manufacture. This was done by putting a molten blob of glass on the body or shoulders of the bottle while the bottle was still hot and pressing a metal seal into it before it hardened. The earliest of these 'sealed bottles' as they are now known, carried seals impressed with coats-of-arms and other heraldic devices. Later, as glass bottles became more popular, initials, surnames, and even complete sentences were impressed on the seals. Samuel Pepys recorded in his diary that on 23 October 1663 he went 'to Mr. Rawlinson's and saw my new bottles, [each] made with my crest upon it, filled with wine, about five or six dozens of them.'

4. The worker on the left uses a special tool to form a kick-up. The worker on the right is rolling the bottle to flatten its sides.

5. Re-heating the neck before working it to final shape. The bottle is being held there on the blowpipe. (*continued*)

6. Using calipers to finish the lip.

It is not known whether Samuel Pepys' bottles included their date of manufacture within the seals, but the inclusion of dates – some of them years of manufacture, others vintage years – was fairly common practice by the end of the seventeenth century. The earliest dated specimen so far discovered in Britain is now in the Guildhall Museum, Northampton, England. It bears the initials RPM, a head thought to be that of Charles 1, and the date 1657. An earlier dated example was thought to have been discovered twenty or thirty years ago during rebuilding work at Compton Castle, South Devon. It was said to bear the date 1620; but the bottle was recently inspected and reported to be *undated*. (*Note:* I have illustrated approximately five hundred seals used on bottles in my earlier book, *Collectors Guide To Seals, Case Gins And Bitters*, Latimer New Dimensions, London, 1975.)

Several factors led to a more widespread use of glass bottles in Britain during the second half of the seventeenth century. Of considerable importance were the ending of glassmaking monopolies by the Cromwellian administration after the Civil War, and the development of coal-fired furnaces which freed glassmakers from the need to have vast forests close at hand. By 1695 the number of glasshouses operating in England had risen to ninety, of which thirty-eight were bottlemakers with a total annual production of more than three million bottles.

It was shortly after that year, probably at the dawn of the new century, that English bottlemakers evolved a method of blowing wine bottles in earthen moulds. These were simple one-piece cup moulds which formed cylindrical sides on bottles blown in them and considerably reduced the time it took to make each bottle. Figs 3, 4, 5, and 6, which show bottlemaking scenes in a French glasshouse in 1750, suggest that the cylindrical shape was adopted on the Continent soon after its development in England.

However, it was in France at some time between 1670 and 1715 that a development of even greater importance than the English cup mould took place. All the bottles discussed so far, from Roman amphorae to English cylindrical wines, were stoppered with loose-fitting corks that had to be laboriously tied and waxed to ensure an air-tight closure and prevent the growth of fungi which turned wine to vinegar inside the bottle. It was a monk named Dom Perignon, manager of the cellars of the abbey of Haut Villers in what is now the Champagne region of France, who successfully experimented with tight-fitting corks between 1670 and 1715 and who produced the first sparkling champagne. Because of the powerful effervescent forces generated inside champagne bottles it was still necessary to wire down corks on these bottles after filling (as it is today), but wiring became unnecessary for other wines when corks could be hammered flush with mouths. This led to a gradual disappearance after 1700 of bottles with string rings. Instead, lips were made much thicker, and therefore stronger, to withstand the blows necessary during flush corking. The following is a description of champagne bottling written in France in 1851. It is reasonable to assume that non-effervescent wines could be bottled and corked even quicker because stringing and wiring were unnecessary.

'The bottles which have been approved are well washed, inverted and allowed to drain, rinsed with spirits of wine, and kept closed with an old cork until required for filling. They are filled to a height of about two inches from the top of the neck. The filler passes it by his right hand side to the corker, who sits on a stool, having before him a little table covered with sheet lead and no higher than his knees. He, judging the space necessary to be left between the cork and the wine, regulates it very nicely, chooses a cork, moistens it, introduces it into the bottle and strikes it two or three times with a wooden mallet so sharply that a stranger might think the bottle must be broken by the sharpness of the blows. Breakage however is rare in the hands of an experienced workman who pays attention to placing the bottle solidly and resting it with a perfectly even pressure upon its bottom. The bottle thus corked is passed again by the right hand to a third workman who ties down the cork with string and then passes the bottle to a fourth who wires it and places each bottle on the ground to form a parallelogram so that the bottles can be counted readily. The men cork, string, and wire one thousand seven hundred bottles per day.'

There can be little doubt that cylindrical wine bottles with lips adapted to take flush corks were being made in England in the early years of the eighteenth century. The need for such bottles became even greater after 1703. In that year the Methuen Treaty between Portugal and England was concluded. Under its terms it was agreed that Portuguese wines should be allowed into England at a lower duty than was then being levied on French wines and that in return England should export greater quantities of woollen and cotton goods to Portugal. As a result Portuguese wines, especially port, achieved widespread popularity in England. The port was shipped in butts and bottled by English vintners who required for this purpose bottles that could be tightly corked and stored in the minimum space during the maturing period so essential to the production of the wine. When laid on its side a bottle of cylindrical body shape ensures that the liquid it contains remains in contact with the cork which in turn does not dry out and shrivel.

Of course, a bottle with a flush cork required the use of a corkscrew

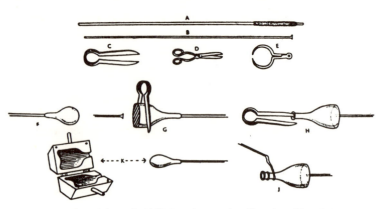

7. Glassmaker's tools in use in 1817. Note the two-piece hinged mould on the left.

to open it. The tool must have been in common use throughout most of the eighteenth century because the first British patent relating to it – Samuel Henshall's of 1795 – refers in the specification to a method of '*improving* corkscrews'.

Throughout the eighteenth century wine bottles became progressively narrower and more straight-sided. At some time before 1817 earthen cup moulds were superseded by hinged iron moulds made in two parts, as seen in Fig. 7, which shows glassblowers' tools used in 1817. Wooden moulds were also in use, as is proved by an invention patented in London as late as 1857 by E. Breffit which described the impregnating 'of wooden moulds and push-ups used in the blowing of glass bottles' with solutions of alum and other sodium silicates 'to render them less inflammable'. Of course, all the bottles so far described, whether formed in earthen cup moulds or in two-piece iron or wooden moulds, required the use of a pontil rod to enable the neck and lip of the bottle to be formed with handtools after removal from the mould. The pontil rod left a scar on the base of the bottle. When a solid rod was employed the scar took the form of a *solid* circle of glass. As is shown in Fig. 5 the bottle could also be held during the working of the lip on the blowpipe which had formed the body. After blowing the bottle was rested on its side and the blowpipe removed by wetting the hot glass close to the end of the blowpipe until the neck cracked, or sheared, from the pipe. The blowpipe was then reheated in the furnace and dipped into molten glass prior to its attachment to the base of the bottle. When removed after the formation of the bottle's lip it left a circular, *open* pontil scar. Both methods were in use simultaneously; which method a particular glasshouse used probably depended on whether or not the glassblower worked alone or with the help of an apprentice when making bottles.

The mention of 'wooden push-ups' in Breffit's invention of 1857 is most interesting. If a wooden push-up was also used to *hold* the bottle during working of the lip it might explain the existence of some early nineteenth century bottles which, instead of the usual pontil scar, reveal bases coated with a chemical substance thought to have been used to attach pontil rods to bottles. The name 'graphite pontil' is used by collectors to describe the mark because chemical analysis indicates that graphite was the substance employed. Unfortunately Breffit's specification does not include illustrations showing the tools so the problem remains unresolved, though it is worth noting that most American bottles with 'graphite pontils' can be ascribed to the period 1840–60.

The next great leap forward in bottlemaking craftsmanship occurred in 1821 when Henry Ricketts of Bristol patented his 'invention for improvements in manufacturing glass bottles'. The importance of this invention cannot be over-emphasised. For the first time it became possible to make bottles of very accurate dimensions and capacities; the need for which became increasingly urgent with the development of bottle filling machines; of more sophisticated packaging and retailing methods; of export markets also, and, too, of the intensely commercial world of the nineteenth century. Ricketts' own words, taken from the text of the original specification, stress these points and reveal some interesting information about bottles of that period.

'My invention comprises an improvement upon the construction of all moulds heretofore used in the manufacture of bottles, whether

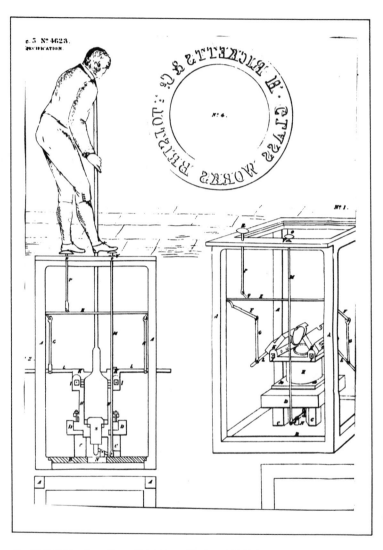

8. An illustration from Henry Ricketts' specification of 1821 showing his improved mould. The metal which embossed the base of the bottle is shown at the top of the drawing.

of black or other descriptions of glass, of which bottles can be made by means of an entirely new method in the construction and operative movements and of appendages of such moulds, particularly in reference to the casting and making of bottles, such as are used to contain wine, beer, porter, cyder, or other liquids. By this my sole invention, the circumference and diameter of bottles are formed nearly cylindrical, and their height determined so as to contain given quantities or proportions of a wine or beer gallon measure, with a great degree of regularity and conformity to each other, and all the bottles so made by me after this method present a superior neatness of appearance and regularity of shape for convenient and safe stowage, which cannot by other means be so well attained.'

The plans which accompanied the specification are shown in Fig. 8. Explaining the workings of his new mould Ricketts went on to say –

'Of the drawing, No.2 is a section representing a square frame of iron, wood, or other material, the legs of which marked A,A, are collateral, and which is fixed in a pit formed in the floor, or placed in any more convenient part of the glasshouse; B, the bottom or floor of the said square frame is sunk or fitted into the grooves or rabbets formed to receive it in the connecting sides at the bottom of the frame. In or near its centre there is a hole to give room for operation of the knocker-up, marked N. The platform or stand D,D, contains a hole in its centre for insertion of the bottom of the mould, so that the knocker-up N may act in projection of the punty or pricker-up S against the bottom of the bottle; E marks an horizontal spindle with a bearing at each end working in two eyes, which are secured to the near legs of the frame A,A, in such a manner as to admit of forming an axle or spindle; at each end of this axle or spindle is an arm marked F,F, projecting towards the centre of the frame and secured by bolts to two other arms marked G,G. These at their ends are formed to receive screws, which admit of being varied in length according to the height or bulk of the mould affixed to the platform D; H represents the body of the bottle mould, of whatever shape or size, placed on its bed D, the lower part of which, as already stated, is open to the operation of the punty S. The mould can be removed at pleasure, according to the size of the bottle required to be made; the bottom or lower part of all the moulds being made to fit the same frame; I,I, bolts of the joints that connect the body and the cover of the mould, and which are fastened with screws or nuts; K,K, the cover or upper part of the mould, which is in two parts, and so shaped that on being closed they form the shoulder and part of the neck of the bottle; L,L, the arms of the cover having a hole near each end, through which small bolts connect them with the arms G,G, above described; M, a rod passing through a hole in the upper part of the frame, and connected by a bolt to one end of a short lever, which works on a round pin near its centre, and at the extremity of the other other end is the knocker-up N. The act of treading upon the mushroom-shaped cap of M, marked O, so raises the knocker-up N against the punty S under the mould, as to produce the concavity usually formed at the bottom of the bottle, and which by this my invention effectually secures a symmetry of shape. P marks another rod passing through a second hole in the upper part of the frame on the same side with M, and connected by a bolt with a short arm marked Q, projecting from the horizontal spindle E. By treading on its mushroom-shaped cap R, Q is pressed down, and raises the arms F,F,G,G, so as to close the cover or upper part of the mould and form the shoulder and part of the neck of the bottle. The foot being removed from the mushroom-shaped cap O, the knocker-up drops into the hole in the bottom or floor of the frame and disengages the punty, and on the foot being taken from the mushroom-shaped cap R, the weight of the arms and handles opens the cover, and the mould is thereby ready for continued working.

No.4 represents a ring or washer placed at pleasure within the bottom of the mould; according to the thickness or thinness of the said ring is the body of the mould shortened or increased, and the various sizes of bottles produced. Upon the surface of this ring can be engraven the address of the manufacture, together with figures or marks indicating the size of the bottle, and which exhibit the same by a projection of the characters in the glass. V a washer, one or more of which may be placed under the mushroom-shaped cap O to decrease the effect of the punty, as may be necessary to regulate the same.'

As he explained in his specification, Ricketts' bottles were moulded

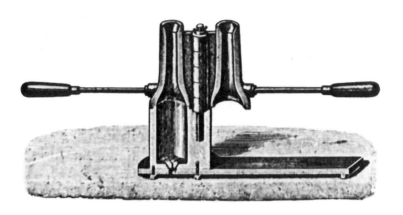

9. Two-piece hinged moulds made from brass in use throughout Europe in the nineteenth century. Metal plates carrying letters to be embossed on the body of a bottle could be inserted before the bottle was blown. (*See also page 27.*)

in three parts – a body and two shoulder and neck pieces. The moulding process left three distinct mould marks on the bottle, one encircling the body beneath the shoulders, the other two vertical on either side of shoulders and neck. These marks, together with the base embossing shown in Fig. 8 are seen on all of Henry Ricketts' bottles made between 1821 and 1853 when Ricketts' Glassworks amalgamated with Powell & Filer, another Bristol manufacturer.

The existence of other bottles carrying dated seals of the early nineteenth century and displaying three-piece mould seams but *without* base embossing suggests that three-piece moulds might have been in use before Ricketts patented his mould. If that was the case it is strange that no British invention for a three-piece mould was patented before 1821. Perhaps the unembossed specimens were made by Ricketts before 1821 during the years he must have taken to perfect his patented invention. Alternatively, it is possible that present-day collectors have been misled by the dates on the seals of these bottles which might be vintage years impressed on the seals of bottles made much later.

Although embossing (*i.e.:* the moulding of company names, product names, and trade marks on a bottle during manufacture) is unlikely to be found on the bottom of bottles made before 1821, it was certainly used on the bodies of bottles made before that date, and probably before the end of the eighteenth century, in two-piece hinged moulds. Letters or words were first cast in brass then placed inside the mould so that the molten glass made an impression of them as the body of the bottle was blown to the shape of the mould. The advantage of embossing over glass seals was that it enabled a manufacturer or merchant to use the entire body area of the bottle to advertise his company or product. Most wine and spirit merchants favoured seals (together with paper labels) during the first half of the nineteenth century, but they had many of their bottles embossed between 1850 and 1870 (Figs. 9 and 10).

Detailed though Henry Ricketts' drawings and specification are, they do not tell us how the bottle was held so that the glassblower could hand-finish the lip after body, shoulders, and neck had been blown. Ricketts' bottles do not have pontil scars and some have embossed letters (usually giving the bottle's capacity) at the centres of their bases which confirm that some other method of holding the bottle must have been used. By the late nineteenth century bottles were being made in semi-automatic moulding machines and the device used then to hold the bottle as it was removed from the mould so that its lip could be machined was called a 'snap case'. It consisted of two semi-circular adjustable clamps held on a rod and it gripped the bottle around its middle, thus obviating the need for a pontil rod (Fig. 11). Perhaps Ricketts used a similar tool. A study of the bases of dated sealed bottles suggests that snap cases *were* in use during the first half of the nineteenth century, but no invention relating to snap cases was patented in Britain. This might indicate that it was a commonly used tool throughout the bottlemaking industry, but it is worthy of note that as late as 1852 F. J. Beltzung patented in London a machine which formed an external screw thread on the lip of a bottle after it was blown in a mould, and that the bottle was held during the process on a pontil rod (See Fig. 12).

Beltzung thus described the workings of his invention.

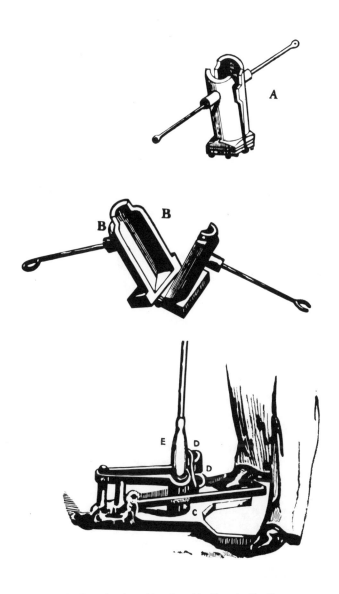

10. Two-piece brass hinged moulds. (*See also Fig. 9.*)

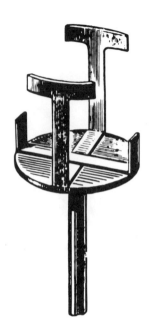

11. Type of snap case in use in France in 1860. Similar tools were probably used in Britain and other European countries.

'U is an arm or bracket fixed to a pillar T and carrying two grooved rollers, u,u, which serve to support the iron rod or punty V on the end of which is fixed the bottle W. The glass bottle is brought in a heated and plastic state from the furnace by means of the punty V, which is laid upon the rollers u,u, and pushed up to the plate or ring H, the neck of the bottle clipping over the closed mandril. The moment the bottle is in its place the workman puts his foot on the treadle, and thus closes the dies C,C1 upon it. Any excess of glass which may have collected on the neck is cut off by the sharp edges of the dies. The workman, or another workman or child, now turns the handle O, by which means the mandril is turned round inside the neck of the bottle, and at the same time allowed to expand by the withdrawal of the socket G. The glass is thus forced into the dies and forms a very perfect screw on the exterior, while the interior is smoothed and

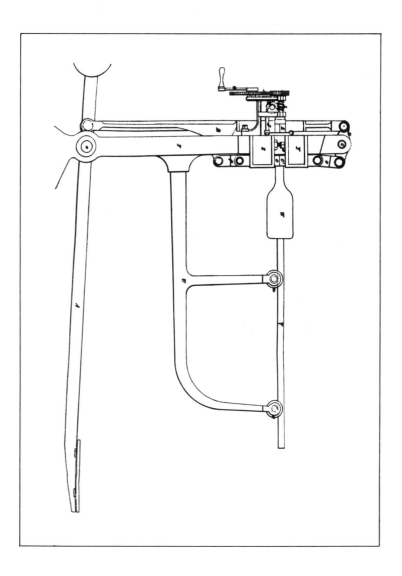

12. F. J. Beltzung's invention of 1852 which clearly shows a bottle held on a pontil rod (v).

rounded by the mandril. As the motion of the handle O is continued the revolution of the wheel K returns the socket G to its first position and thus closes the mandril. The foot is then taken off the treadle, and the bottle drawn back off the mandril, and carried away, and the operation is completed.'

(*Note:* The inventor suggested the use of pewter screw caps with these bottles.)

The introduction of labelling machines in the bottling industry after 1870 induced the majority of wine and spirit merchants to use colourful paper labels on their bottles, though a substantial minority continued to use sealed or embossed bottles until the 1890s.' The ideal bottle for use with a paper label was one without mould seams on its body. When labels were hand-applied the labeller simply turned the bottle so that the label adhered to a part of the body that did not have a vertical seam which might have spoiled the effect of the label. But machines could not do this; they simply attached a label on to whatever part of the bottle presented itself as it moved along the conveyor belt. It is for this reason that bottles made in three-piece moulds were widely used for wines and spirits. Dozens of inventions for three-piece moulding machines were patented in Britain between 1860 and 1880, but 1879 saw the invention of a technique of moulding bottles *without* seam marks. This was patented by G. Evinson in April of that year and the specification described a method 'of avoiding mould marks by lining moulds with a mixture of plumbago and oil or tallow by which combustion produces gas and forms a cushion enabling the bottles to be easily rotated in the mould'. The invention also described a method 'of embossing the bottoms of turn-moulded bottles'. This was achieved by inserting into the bottom of the mould a loose-fitting metal ring carrying the letters to be embossed. The ring turned *with* the bottle as the latter was blown and rotated in the mould.

The evolution of wine bottles is thought to have followed a similar pattern throughout Europe. By the 1880s most French, Spanish, Portuguese, Italian, and German wine merchants had abandoned seals and embossing in favour of turn-moulds with paper labels. Wine bottles had by then become so cheap to manufacture they were used as 'one trip' containers not required to be returned by the customer. Therein lies another reason why wine merchants used paper labels on plain, turn-moulded bottles; unlike mineral water bottles which were frequently returned, washed, and refilled, wine bottles were used only once and could therefore be identified by a 'mark' which would certainly have fallen off during washing.

Excavations on colonial sites in America and Australia have established that English-made sealed bottles were widely used in the British colonies. More than 20,000 fragments of English-made glass bottles, including 106 seals, were recovered during excavations on the site of Jamestown, Virginia, which was established by English colonists in 1606. Australia remained totally dependent on Britain for glass bottles until the late nineteenth century, but in America a glass-works was established in 1739 at Salem County, New Jersey, by Caspar Wistar. Another successful glassworks established before America's independence was that of Henry William Stiegel which commenced operations at Manheim, Pennsylvania, in 1763. Both Wistar and Stiegel produced copies of contemporary European glassware which must surely have included sealed wine bottles.

By the beginning of the nineteenth century glass was the most widely used packaging material in the Old and New Worlds. Its cheapness certainly contributed to its popularity, but it was the proliferation of products suitable for packaging in bottles which gave glassmakers their vast markets and thus enabled them to make bottles very cheaply. The manufacture of wine bottles provided 'bread-and-butter' for European glassworks in the seventeenth century, but bottles for brandy and other distilled spirits were also required in large numbers in the eighteenth century. Alchemists and apothecaries had been experimenting with alcohol and its purification since the twelfth century and brandy was known and drunk throughout Europe by 1400. (It was prescribed by physicians during the plague of the Black Death in 1348–52.) The herb gardens of medieval monasteries were the birthplaces of liqueurs such as Chartreuse and Benedictine which soon achieved widespread popularity and which became serious competitors to wine and beer in the nineteenth century.

Although early production of spirits was from wine its successful production from fermented cereals was achieved as early as the fifteenth century. This was most important to beer-producing countries which had to import the wine from which they made brandy. Gin, distilled from grain and flavoured with juniper berries, was introduced into Holland in the sixteenth century by German soldiers from Hanover serving in the wars of liberation against Spain. From there it went to England (in square 'case bottles' designed specially as export containers) where in the early eighteenth century its sale under the sign, 'Drunk for a penny; dead drunk for twopence' was one of the great evils of the times. Its consumption by the upper classes under the name of Schnapps was regarded as more socially acceptable. Whisky, once 'the drink of the rebel Scot', gained the approval of genteel English society in the eighteenth century. By the beginning of Victoria's reign (1837) the following alcoholic drinks were included in the wide range advertised in British and colonial newspapers.

'Port, sherry, Marsala, Madeira, claret, hock, champagne, London gin, West Indian rum, English rum, French brandy, Hollands gin, Scotch malt whisky, Irish whiskey, British brandy, green ginger wine, ginger brandy, Curacao punch, orange brandy, cherry brandy, cherry liqueur, creme-de-Noyeau, Chartreuse, Benedictine, London porter, East India ale, black beer, brown stout, bitter beer, Dublin stout, Burton ale, and Scotch ale.'

A further indication of the remarkable growth of the bottling industry that took place in the nineteenth century can be gained from the amount of cork imported to Britain in 1866 – 6,241 tons, *plus* 2,684,000 ready-made bottle corks. Almost all of it came from Portugal as dunnage in ships ladened with wine.

2 Medicine and pharmaceutical bottles

An equally profitable market for nineteenth century bottles was provided by the sale of pharmaceutical products. Like the production of alcoholic drinks, this industry was born in the laboratories of medieval monks, alchemists, and apothecaries who stored opium, belladonna, and dried herbs in ornate druggists' jars and dispensed their secret nostrums and elixirs in tiny, free-blown vials. Until the seventeenth century these containers were unmarked; the name of the 'chymist', his medicine, and instructions for its use were often printed, together with testimonials from satisfied customers, on a large sheet of paper in which the bottle was wrapped and secured with string and wax. Gradually, and with the help of advertisements in the increasing number of newspapers which began to circulate in the late seventeenth century, some of these cures achieved national reputations. Their inventors, anxious to protect themselves from competitors, began to mark their bottles by impressing seals bearing their names into the wax which covered the cork. Those who sold their products through 'agents' published advertisements warning customers not to accept bottles with 'broken seals' which indicated that bottles had been refilled with 'spurious imitations'. At some time in the eighteenth century the practice grew up of embossing bottles with the name of the chemist and (later) the name of the medicine; but seals impressed into wax-covered corks continued to provide the best defence against 'fraudulent refilling' for many years to come. The following advertisement, published in the *Yorkshire Gazette* in 1847, is typical of thousands which appeared in the late eighteenth and nineteenth centuries.

'Do you want luxuriant hair and whiskers? Rosalie Coupelle's Crinutriar is guaranteed to produce whiskers, moustachios, eyebrows, etc. in a few weeks, and restore the hair in baldness, from whatever cause, strengthen it when weak, prevent it falling off, and effectually check greyness in all its stages. For the nursery it is recommended by upwards of 100 physicians for promoting a fine, healthy head of hair, and averting baldness in after years. Sold by all chemists or sent by post free on receipt of 24 penny stamps by Miss Coupelle, 69 Castle Street, Oxford Street, London. CAUTION: See that the words Coupelle's Crinutriar are moulded in each bottle and that each package is stamped with the name Rosalie Coupelle over the cork.'

Bottle collectors use the name 'patent medicines' to describe all bottles embossed with the names of cures and elixirs; but strictly they ought to be divided into *patent and proprietary medicines*. A patent medicine was protected by Letters Patent; the formula from which it was concocted could not be copied until the patent expired. A proprietary medicine was protected *only* by its trade name; a competitor could make up an identical medicine and sell it under a different name. Of the *thousands* of medicines sold in the eighteenth and nineteenth centuries only a few hundred were patented. The first of these, 'for making of salt of purging water' was patented in London in 1698 by Nehemiah Grew. Those which achieved international fame, the empty bottles having been found in dumps on both sides of the Atlantic and in the southern hemisphere, include Richard Staughton's Elixir (patented in 1712); Robert Turlington's Balsam of Life (1744); and Robert Walker's Jesuit Drops (1755).

Family chemists who made up their own preparations at the backs of their shops and sold them in packages tied with paper and string were overtaken in the nineteenth century by manufacturing and wholesale chemists who sought entire nations as customers for their products. These 'giants of industry and commerce' either put the 'little men' out of business with price-cutting campaigns or gained control of their businesses by persuading them to sell their secret formulae and have their preparations made up and packaged more efficiently and at less cost in a modern factory. Of course, if a particular medicine had already achieved a good reputation the manufacturer or wholesaler retained its original name even though it was no longer made by the inventor. Thus we find in nineteenth century advertisements a rich assortment of names and brands emanating from a single source.

'Customers are reminded that Carpenter's Vegetable Specific (bottles 1/1½d); Marshall's Heal-All (1/1½d); Wray's Concentrated Essence of Jamaica Ginger for nervous complaints (2/9d); Frank's Specific Solution of Copaiba for female complaints (2/9d); Woodhouse's Balsam of Spermacceti for coughs (2/9d); Paul's American Balsam for the chest and lungs (1/1½d); Ramsbottom's Corn and Bunion Solvent (1/1½d); Church's Cough Drops for asthma (2/9d); and True Daffy's Elixir for cholic (2/9d and 3/6d) are all sold wholesale to our agents only by W. Sutton & Co (late Dicey & Sutton), 10 Bow Church Yard, London. Available from our agents throughout Yorkshire. Purchasers are warned not to rely merely upon the glass bottles bearing the name of Dicey & Co, as there are unprincipled people who buy up the empty bottles for the purpose of filling them with their own counterfeit preparations. The only criterion is to examine whether the stamp affixed over the cork has the words Dicey & Co. printed thereon and to observe that the bill of directions is signed W. Sutton & Co., late Dicey & Sutton.'

(*Yorkshire Gazette, 1837.*)

Many of the warnings issued by manufacturers and wholesalers suggest that some of the little men who at first lost business to the giants of the industry learned quickly to benefit from their rivals' expensive advertising by selling 'spurious imitations' of the branded products with which the public soon became familiar.

'Evan Edwards, sole manufacturer of Evan's Pectoral Balsam of Honey for coughs and colds (bottles 2/9d and 3/6d) reminds the public that no medicine was ever so universally counterfeited as this, and a few years ago the wife of the Rev. Robert Fowler of Peterborough fell heavy sacrifice to a spurious Balsam of Honey sold by a chemist and druggist in London. It is necessary, therefore, to notice that the genuine preparation has engraved on the stamp over the label "Evan Edwards, 67 St. Pauls, London".'

(*Yorkshire Gazette, 1837.*)

And –

'Rowland & Son, makers of Rowland's Macassar Oil and Rowland's Kalydor for eradication of eruptions and pimples warn the public that many shopkeepers offer for sale counterfeits of the above, composed of the most pernicious ingredients. They call their trash the "GENUINE" and sign "A. Rowlandson" – omitting the &, recommending them as being cheap. Be sure to ask for ROWLAND'S by name.'

(*Yorkshire Gazette, 1837.*)

In addition to providing a gullible public with numerous cures, elixirs, unctions, specifics, balms, and heal-alls, nineteenth century chemists also sold a vast range of 'druggist's sundries'. These included perfumes, pommades, smelling salts, disinfectants, poisons, and many other products packaged in glass. The finest smelling salts cost, in 1837, up to ten shillings per ounce, while perfumes commanded even higher prices. Their sale in cheap and crudely-made bottles would have been difficult if not impossible; it was essential that the container looked as expensive as its contents. Consequently glassworks were deluged with orders for intricately moulded bottles in glass of the best quality and in a wide range of colours (See Fig. 13). Poison bottles and display bottles for chemists' shelves were also required in large quantities. The business of supplying them to retail chemists became so lucrative that many manufacturing and wholesale chemists either took over or bought substantial interests in glassworks and devoted production entirely to the manufacture of containers for the pharmaceutical industry. Figs. 14 and 15, pages from a British wholesale chemist's mid-nineteenth century catalogue, also give an indication of the range of bottles likely to have been displayed on the shelves of a chemist's shop of that period.

---

13. A page from a nineteenth century wholesale chemist's catalogue showing his range of smelling salts bottles.
(*Facing page.*)

14., 15. Pages from a British wholesale chemist's mid-nineteenth century catalogue showing display bottles and other wares available to retail chemists at that time.

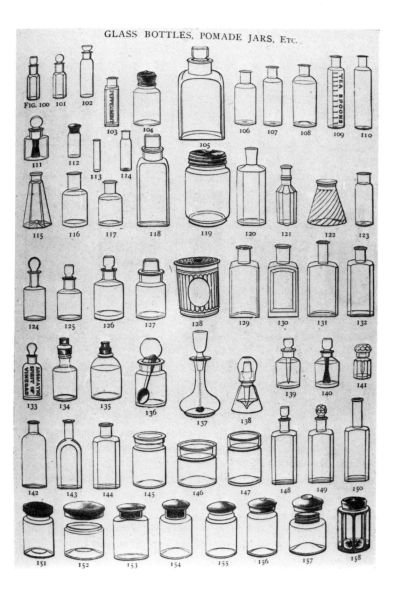

In the United States a similar sequence of events took place as the sale of patent medicines grew from humble colonial beginnings to a major nineteenth century industry. In the eighteenth century all the elixirs and cures on sale were imported from Britain, but after gaining independence America soon became almost self-sufficient in quacks and quackeries. Indeed, the quantities of American medicines exported to Europe in the nineteenth century was probably greater than the imports of the previous century. The phenomenal growth of the industry can be illustrated by sketching the career of Thomas W. Dyott, one of America's best-known purveyors of quackery. Many of his competitors and imitators enjoyed similar success.

Thomas W. Dyott emigrated from Britain and settled in Philadelphia in 1805. He was then twenty-eight years old and possessed, in addition to a burning desire to make his fortune, a number of formulae for patent medicines which he had probably 'borrowed' from former employers on the other side of the Atlantic. (One of his recipes was for a non-mercurial medicine for venereal complaints, a cure widely advertised in British newspapers at that time.) His first business venture in the New World was as a purveyor of boot blacking; but by 1807 he had established a 'patent medicine warehouse' from which he sold a range of products claimed to be made to the formulae of 'Dr. Robertson'. It seems likely that these medicines were imported from Britain. They either arrived in Philadelphia already bottled or they were 'put up' by Dyott in unembossed bottles with paper labels. However, business was so good that by 1809 he could afford to have his bottles specially made. An advertisement published that year stated that his medicines were 'packaged in American-made square flint glass bottles blown in Doctor Dyott's private mould in which the inscription "Dr. Robertson's Family Medicines. Prepared only by T. W. Dyott" is cut.'

Note that by this time Dyott had added the title 'Doctor' to his name. It seems greatly to have boosted his reputation. Within a few years his advertisements were displayed in almost every newspaper in the eastern states, he had appointed numerous 'agents', and he was also kept busy at his 'private apartments' where he offered 'secret and discreet treatment without mercury to persons having contracted venereal complaints'.

America was at war with Britain between 1812 and 1815, a war which cut off supplies of imported medicines and encouraged Dr. Dyott (and others) to concoct imitations of many British products. Dyott went further. After acquiring an interest in a local glassworks he turned out copies of well-known British medicine bottles and filled them with his own products. These included 'Robert Turlington's Balsam of Life' and 'True Daffy's Elixir'. He also made an assortment of 'vials, preserving jars, and quart and half-gallon bottles' which he sold to retail chemists who could no longer obtain imported supplies. By 1822 he was owner of a substantial glassworks turning out a wide variety of glass bottles. In advertisements published in that year he offered:

'Every description of wares including apothecaries' vials ½ oz–8 oz; patent medicine vials of every description; mustard, cayenne pepper, olives, anchovies, sweet oil, and cologne bottles; scotch, rapee, and maccabow snuff bottles; confectionery and apothecaries' shop bottles; pickling and preserving jars; demi johns; Bateman's vials; British Oil vials; Staughton's vials; Turlington's vials; Peppermint vials; Godfrey's vials; Haarlem Oil vials; Dalby's vials; Opodeldoc vials; Daffy's vials; Cepholic Snuff vials; Lemon Acid vials; Balsam of Honey vials; Jesuit's vials; Whitehead's Essence of Mustard vials; Clarke's ink vials; castor oil bottles, seltzer water, blacking, varnish, and tincture bottles.'

It is known that the fabrication of small vials, whether free-blown or blown for body shape in two-piece moulds, demanded special skills not mastered by all glassblowers. It is also known that substantial numbers of English glassblowers went to America in the early years of the nineteenth century. Some made the journey as emigrants; others were sent during periods of high unemployment on 'loan schemes' organised by British glassworkers' unions. Most of those who crossed the Atlantic on 'loan schemes' came from the Birmingham area where the blowing of fancy bottles was an established industry. They were sent to Philadelphia and New Jersey and it is probable that some of them were employed by Dr. Dyott and helped him achieve a high reputation for the quality of fancy cologne and pungent bottles produced at his glassworks. By 1835 these included.

'Colognes – Hexagonal, Fluted, Long Gothic, Urn, Lion, Dragon, Lyre, Fountain, Round Flowered, Bellows, Barrel, Castle, Square Flower, Column, Flower Basket, Acorn, Cathedral, Diamond, and Lily.

Pungents – Oak Leaf, Diamond, Acorn, Pineapple, American Shield, Urn, Cornucopia, Grapes, Strawberries, Harp, Magnolia, Dolphin, and American Eagle.'

Thomas Dyott eventually made a fortune from his patent medicines and his bottleworks, though he lost it all later and went to prison for bankruptcy. With the use of massive newspaper advertising, confidence trickery, and high-pressure salesmanship his imitators achieved even greater successes. By the 1870's they were marketing at least 40,000 cures and painkillers in a fascinating array of colourful bottles and jars. Total sales were to reach an annual turnover figure of eighty million dollars before 'the great American fraud' was finally exposed in 1905.

3 Mineral water bottles

Another nineteenth century industry which made the fortunes of entrepreneurs and inventors and which also required the production of enormous quantities of special bottles was the manufacture and retailing of artificial mineral water. Again the beginnings can be traced back to medieval Europe where physicians first noted the beneficial effects of drinking naturally mineralized water. The fame of those places rose to the surface as springs spread rapidly and pilgrimages to partake of 'the healing waters' were prescribed for all sorts of ailments and maladies. By 1750 it had become fashionable among the upper classes to take long holidays at these 'watering places' around which were built lavish bathing pools, pumprooms, ballrooms, and hotels. The German resorts of Selters and Baden-Baden achieved international fame, as did Bath, Tunbridge Wells, Cheltenham, and Harrogate in England; at the height of their popularity in the mid-nineteenth century these great spas counted their annual visitors by the million.

There were of course even greater numbers who could not afford the expense of a visit to a spa and it was to serve their needs that the practice grew up of selling *bottled* spa water. The owners of mineral springs held a monopoly on this business until the late eighteenth century when chemists began to analyse the mineral contents of spa waters and to sell palatable imitations at much lower prices than those charged for products 'bottled at the spring'. Pioneers included Dr. Joseph Priestley who in 1772 devised an apparatus for 'aerating' distilled water used on ships of the British Navy, and John Mervin Nooth who in 1775 invented a 'Gazogene' for the manufacture of aerated water in the home. Large scale commercial production of artificially carbonated water was first undertaken in England by Joseph Schweppe who opened a factory in Drury Lane, London, in 1793. His 'soda water and Seltzer water' were widely consumed by the beginning of the new century.

Another pioneer, William Hamilton of Dublin, is usually given credit for the invention of the egg-shaped bottle used in Britain throughout the nineteenth century as a container for soda water. In his patent of 1809 for 'preparing soda and other mineral waters' Hamilton described the bottle thus –

'I generally use a glass or earthen bottle or jar of a long ovate form, for several reasons, viz, not having a square bottom to stand upon, it can only lie on its side, of course, no leakage of air can take place, the liquid matter being always in contact with the stopper. It can be much stronger than a bottle or jar of equal weight made in the usual form, and is therefore better adapted for packing, carriage, etc. The neck of the bottle and the mouth are sometimes formed that it may serve as a drinking glass if necessary. I commonly stop with cork, which, from the excessive pressure generally existing within the bottle, flies out on the detaining strings being cut.'

Most collectors refer to these egg-shaped soda water bottles as 'Hamiltons', though it is possible that bottles of this shape were in use before 1809. A few early examples with pontil scars have been found but the majority were held in a snap case during formation of the thick lip. In the specification accompanying an invention for 'improvements in bottles for aerated liquids' patented in London in 1848 Alfred Hely and Joseph Norton described a method of blowing egg-shaped bottles to give them a greater thickness at their shoulders and necks, 'thus reducing risk of bursting during filling'. They explained the method of blowing as follows.

'A bottle of this character may be manufactured by blowing through and twisting round the rod placed in a vertical position, or nearly so, above the head of the operator previous to inserting the same in the mould and finishing it, but care must be taken that after blowing the metal be allowed to get sufficiently cool to prevent the chance of it flowing back to the punt of the bottle when inserted in the mould; or the chance of this may be prevented by fixing the

16. A drawing from Hely and Norton's specification of 1848 showing the thicker glass at neck and shoulders of a Hamilton bottle achieved by holding the blowpipe above the head so that the red-hot glass flowed downwards as the bottle cooled.

mould overhead, mouth downwards, instead of blowing mouth upwards as at present.' (See Fig. 16.)

Until the 1830's it was the 'medicinal benefits' of artificial mineral waters that recommended them to the public; but in Britain in 1833 they were exempted from the duty levied at that time on medicines. The reduction in price, together with the practice which began at about the same time of adding 'flavourings and syrups', encouraged the public to drink carbonated waters for their refreshing and thirst-quenching properties. Inevitably this led to a massive increase in consumption and a spectacular increase in the number of mineral water makers. A similar and speedier rise in popularity occurred in the United States when 'soda water fountains' offering a wide variety of flavours were introduced in the early years of the nineteenth century.

By the 1870's millions of gallons of lemonade, limeade, cherryade, and other fruit-flavoured aerated beverages were being bottled annually. In Britain entire glassworks were devoted exclusively to the manufacture of mineral water bottles, most of which were by that time stoppered with ingenious internal plugs and balls which relied for their effectiveness on the gas pressure inside the bottle which pressed the stopper into the mouth. Most famous of these bottle factories was the Hope Glassworks at Stairfoot near Barnsley in Yorkshire. Here Ben Rylands made under licence the Codd bottle with its renowned ball stopper and crimped neck. It had been perfected in 1872 by Hiram Codd of Camberwell, who because he lacked the capital to start his own factory, allowed his invention to be made under licence by several manufacturers. Ben Rylands was so successful at making these newfangled bottles that in 1876 he persuaded Hiram Codd to join him in partnership at the Hope Glassworks where, until Ben's death in 1881, they combined inventive genius with manufacturing skill and thereby gained a substantial proportion of the British market in mineral water bottles.

On his father's death Dan Rylands became Codd's partner, but their association lasted only until 1884 when Rylands paid Codd a large sum of money and gained complete control of the business. Like Codd, Dan Rylands also possessed inventive skills and during the next few years he patented a number of improvements to the 'Original' Codd bottle; by 1887 he was able to claim that his company was 'the world's largest manufacturer of mineral water bottles'. (See Fig. 17.) However, he certainly did not have a monopoly on inventions for internally-stoppered bottles and he faced stiff and relentless competition from other inventors and manufacturers who introduced numerous alternatives to Ryland's globe-stoppered bottles throughout the late nineteenth century until the market was eventually lost in the early years of the twentieth century to screw-stoppered bottles. (*Note*: Interested readers will find a complete and fully illustrated catalogue of all internally-stoppered bottles invented between 1868 and 1907 in my earlier book, *Fletcher's Non-Dating Price Guide to Bottles, Pipes, and Dolls' Heads*, Blandford Press, 1976.)

The theft and re-use of bottles by rival manufacturers was a constant problem for nineteenth century mineral water makers. In 1880 one could buy for less than £50 all the plant and machinery required to make artificial mineral water. Thousands of under-capitalized backstreet firms were established and they provided a lucrative market for bottle thieves, most of whom were scavengers

17. Illustrations from Dan Rylands' catalogue of 1889 showing his four best-selling globe-stoppered bottles – the Valve; the Reliance; the Acme; and the Codd Original.

employed by local corporations to empty dustbins. Although customers were expected to return empty mineral water bottles to retailers, deposits were rarely charged until the late 1880's so the scavengers were able to make handsome second incomes by recovering throwaways and selling them to small manufacturers at half the price charged by glassworks for new bottles. What made matters worse was that all mineral water bottles were embossed with company names and trade marks because the use of paper labels was impractical on bottles that were washed and refilled many times. Thus could the reputation of a manufacturer of high-quality mineral water be jeopardized by backstreet firms who refilled his bottles with inferior beverages.

Several methods of dealing with the problem were tried in the 1880's. Stiffer penalties were meted out to scavengers caught removing bottles from dustbins, and agreements were made between the larger mineral water makers to return to their rightful owners any stray bottles that found their way into the crates of another company.

47

Some firms also introduced deposit charges ranging from one farthing to two pence, but this extra payment was often resented by customers who went out of their way to find brands that did not carry deposits on their bottles. In 1887 Dan Rylands patented a method of marking his globe-stoppered bottles in such a way that they could easily be spotted even when placed in the centre of a crateful of bottles owned by a rival. For an extra charge of a few pence per gross Rylands' customers could buy from the Hope Glassworks bottles with lips made from brightly coloured glass. Rylands undertook to supply only one mineral water maker in each town with bottles having lips of a particular colour. It was an innovation that worked well in country districts and in the colonies of Australia, New Zealand, and South Africa where there were fewer mineral water makers. But in major cities with hundreds of mineral water firms which were constantly amalgamating and extending their territories coloured lips were less effective as a protection against bottle thefts. Nevertheless, Rylands' competitors found it necessary after 1887 to offer their customers a choice of different coloured glass when ordering new bottles. Because Dan Rylands held patent protection on the idea of colouring only the lips of bottles his competitors were obliged to make their 'anti-theft' bottles entirely from coloured glass. It is these delightful specimens in amber, dark green, and blue glass which are most prized by mineral water bottle collectors in Britain and Australia.

Although enormous quantities of mineral water were bottled and sold in the United States in the late nineteenth century the legacy of empties found by American dump diggers lacks the variety and colour of those found in Britain, Australia, and South Africa. The British Codd bottle and its numerous rivals failed to win large markets in America where the vast majority of mineral water makers favoured the Hutchinson bottle, an American invention which had a disc-shaped rubber stopper inserted in its neck which could be pulled to a closed position by a metal hook protruding from the bottle's mouth. It was used throughout the United States until the introduction of crown-capped bottles in the early years of the twentieth century.

4 Household bottles and jars

The blowing of wine, spirit, beer, and mineral water bottles and the moulding of glass containers for medicines and druggists' sundries did not account for all production in European and American glassworks during the nineteenth century. There were other markets which also required glass bottles and jars by the million, notably the pickling, preserving, and food packaging industries which evolved as rural populations crowded into cities to tend machines instead of cows, thereby creating a need for packaged foods that could be stored for long periods. In medieval Europe food could only be preserved by salting, smoking, or drying. Most people were dependent during winter months on salt pork and smoked 'red herrings', though noblemen were able to supplement this diet with freshly killed venison and fowl. For those who could afford the luxury the boredom of these fish and fowl dishes might occasionally be relieved by the addition during cooking of rare spices which found their way to Europe via tortuous overland camel and pack-horse routes from the 'mysterious Orient'. It was not until the eighteenth century that the empire-building activities of Britain, France, and Holland brought spices to northern Europe in sufficient quantities to reduce prices to a level at which common folk could enjoy them. Britain's East India Company sent spices by the shipload to the mother country, and soldiers and colonists who returned home after periods of duty in India brought with them insatiable appetites for the chutneys and sauces they had enjoyed in Calcutta and Bombay. By the beginning of Victoria's reign these could be purchased, together with exotic foods from other parts of the world, at the larger grocery stores in most towns.

'William Tonge, Foreign Provision Merchant, Seale Lane, Hull, has just received Indian Curry Powder, Bombay Chutney, Indian Soy, Chinese Preserved Ginger, and East and West Indian Pickle. W. Tonge wishes to remind the public he still prepares his Essence of Gorgona Anchovy, Harvey's Sauce, Reading's Sauce, Emperor of China's Sauce, and Japan Sauce which combine all that is necessary for a Gentleman's table.'

(*Yorkshire Gazette, 1837*)

By the 1850's there were at least one thousand brands of pickle, sauce, and bottled fruit on sale in Britain and firms including Crosse & Blackwell, Lea & Perrin, Lazenby, Keiller, Coleman, Slea, and Goodall & Backhouse were selling their products to large home and colonial markets. To reduce packaging costs many small firms ordered standardized bottles and jars without embossing. These could be made in large batches at the glassworks and sold in small lots to many companies. They were mostly aqua in colour and blown in simple two-piece moulds. On narrow bottles necks were separated from the blowpipe by applying a few drops of water until the glass sheared. This left a jagged top which bit into the cork and thereby provided a suitable closure for a sauce bottle. Wider mouthed bottles and jars used for fruits and jams had their necks strengthened by a ring of glass around the lip so that the neck would withstand the heavy blows necessary to drive home a thick and tight-fitting cork. Bottles used to contain olive and castor oils were usually made in dark blue (cobalt) glass to reduce passage of light which might otherwise have rendered the oil rancid. More expensive versions of these blue bottles (used for the export market) were often blown in decorative moulds which produced bottles with embossed spirals and other abstract designs on their surfaces.

In rural America and Australia many housewives cut off from shops and stores selling proprietary brands were encouraged to make their own pickles and preserves after 1858 when an American inventor, John Landis Mason, patented a simple but efficient screw-topped preserving jar for use in the home. The earliest of these had zinc lids;

later versions had lids lined with porcelain to prevent corrosion when the lid came in contact with the acid in fruit juices. They sold widely in the United States and in the British colonies, but were less popular in England.

'A word in season. Careful housewives should now prepare a bountiful store of food for winter use. Cherries, gooseberries, and vegetables can now be preserved very simply in the new American preserving jars with porcelain-lined screw tops and will keep for any time. Quart jars 10/– per dozen; two-quart jars 12/– per dozen from S. Hebblewhite, George St., Sydney.'

(*Sydney Morning Herald, 1880*)

5   *Ink bottles*

Steel-nibbed pens had already begun to replace old-fashioned quills when the Penny Post was introduced in Britain in 1840. By 1842 Joseph Gillot, pen manufacturer to Queen Victoria, was turning out 70,000,000 steel-nibbed pens per year at his Birmingham factory and a year later production reached 105,000,000 to keep pace with the growing popularity of letter writing. The rise in demand for ink was equally spectacular. A patent for the manufacture of black ink from 'a powder mixed with water, beer, ale or wine' was granted in 1688 to Charles Holman. It was in this powdered form that most ink was sold throughout the eighteenth century, though ornate and expensive glass inkstands and inkwells in which one mixed one's own ink were on sale at that time. But it was not until letter writing had become a widespread habit that liquid ink was sold by the bottle. One of the first vendors in Britain was Henry Stephens of London. He was selling liquid ink by the pint in stoneware bottles in the 1830s, and in the 1840s he advertised its sale in 'penny glass bottles'.

'Stephens Writing Fluid comprises the most splendid and durable colours, and the most indelible compositions which art can produce; they contain the fullest proportions of colouring matter and those to whom economy is more an object than powerful contrast with the paper may dilute them with rainwater to the extent of colour they may require, and thus effect a real economy without paying for the cost and incumbrance of large bottles. One penny glass bottles may be had from all stationers and booksellers. Henry Stephens, 54 Stamford Street, Blackfriars Road, London.'

(*Morning Chronicle, 1842*)

In order to sell the product at a retail price of one penny Stephens, and his many competitors, had to package it in the cheapest possible container. This proved to be a small glass bottle made in a two-piece mould. Its neck was sheared from the blowpipe to leave a jagged rim which bit into the cork to provide a leak-proof closure. Glassworks were obliged to make them by the million in order to keep the price low; for this reason very few were embossed. However, by the use of different moulds it was possible to produce a wide range of shapes and give ink makers some choice when deciding on containers for their latest products. An eight-sided bottle proved most popular, but squares and oblongs were almost as widely used. Occasionally, perhaps when marketing a slightly more expensive brand of ink, manufacturers used figural bottles moulded in the shapes of birdcages, barrels, cottages and other forms. The use of these ornate ink bottles was more widespread in the United States than in Britain, as is proved by the delightful specimens found by American dump diggers.

6 *Figural bottles*

Figural bottles were used in the nineteenth century by manufacturers of other products when the cost of a special mould could be justified by the occasion or by an increase in the price of the bottle's contents. A few British wine and spirit merchants ordered bottles moulded in the form of a bust of Queen Victoria in the Jubilee Year of 1887 and other important events such as the winning of a war or the successes of a famous politician were sometimes commemorated by the use of special glass bottles. One or two specimens from France and other European countries indicate that they were also used on the Continent; but it was in the United States that historical events were most often marked by the production of special glass containers. During the first half of the nineteenth century American whisky firms used millions of bottles embossed with patriotic emblems and the busts of leaders of the nation. Dr. Dyott, whose glassworks has already been mentioned, offered in his advertisements for 1822 a range of these 'historical flasks' including specimens embossed with the American Eagle and the bust of Franklin. In the 1840s some American ink makers sold their products in bottles in the shape of log cabins to commemorate the 'Log Cabin and Hard Cider' election campaign of William Henry Harrison and John Tyler. Others issued ink bottles in the shape of locomotives during the transcontinental railway-building era, and a number of quack medicine vendors boosted their sales by offering their cures in bottles moulded in the shape of Indians, American soldiers, and the type of cannon used in the American Civil War. Enthusiasm for figural shapes in glass bottles lingers to this day in the United States where manufacturers of cosmetics are still able to increase their sales by offering after-shave lotions and perfumes in bottles moulded in the shape of John F. Kennedy's bust, Apollo spacecraft, and other eye-catching designs.

In Britain a small number of figurals were issued by Victorian confectionery makers who sold lemonade powder and other children's delights in bottles in the shape of tin soldiers. One or two lamp manufacturers copied candle-holder shapes in the glass bodies of their lamps; but the only glass figurals produced in large numbers in the nineteenth century in Britain were the basket-shaped coloured glass fairylights which adorned Victorian Christmas trees. A small candle was placed inside each basket and the 'lights' were then hung on the tree by means of thin wire handles fitted around their necks. They dwindled out of use in the early years of the twentieth century when electric lights for Christmas trees came on the market.

7 Coloured glass in bottlemaking

The use of black glass for early wine and porter bottles was not a deliberate choice of colour by bottlemakers; the blackness was produced by impurities, including iron oxide, present in the sand used to make the glass. Wine merchants and brewers were quite happy to use black bottles because the dark glass hid unsightly sediments from the eyes of customers. In fact they encouraged bottlemakers to add more iron oxide to the crucible during melting in order to make glass of even greater opaqueness. But truly opaque black glass was never achieved by nineteenth century bottlemakers; hold any 'black' wine bottles up to strong light and the glass will be seen to be either dark brown or dark green.

The almost complete lack of patents relating to the manufacture of coloured glass suggests that the methods of producing most colours were common knowledge in the industry from an early date. Only the manufacture of red glass seems to have caught the interest of inventors, the earliest British patent being that of Mayer Oppenheim in 1755 which specified copper oxide and a high kiln temperature for the production of 'transparent red glass'. A well-known method of producing ruby red glass was by the addition of small amounts of gold to the crucible during the melting stage, though such expensive glass was not used for making common bottles.

The use of blue, dark green, and dark brown glass as methods of identifying poison bottles had become accepted conventions throughout Europe and North America by 1870. Earlier poison bottles were made in pale green glass, as described in a patent granted to T. Richardson in London in 1855 for making 'pale glass suitable for medicine bottles' by adding borate of lime to the glass mix. The widespread adoption of blue, dark green, and dark brown poison bottles coincided with the general introduction of another method of identifying these bottles, namely the embossing of raised patterns on their surfaces so that they could be recognised by touch in darkness, a necessary precaution in the days before electric lighting. Non-poisonous medicines, especially the quack cures sold widely in America in the nineteenth century, were most commonly contained in pale brown or amber bottles, though a few vendors favoured smokey olive green. Aqua glass was more widely used for these bottles after 1880.

Aqua was also the common colour for British mineral water bottles until the 1880s when Dan Rylands introduced coloured lips – blue, dark green, red, and amber – on his anti-theft bottles. The blue glass was of a pale shade and made by the addition of copper oxide to the crucible, unlike the dark blue of poison bottles which was achieved by adding cobalt to the glass mix. Some of the mineral water bottle makers who competed with Dan Rylands by producing totally coloured bottles used cobalt to make the glass for their blue bottles; others produced paler copper-blue specimens. The range of browns extended from pale honey-amber to deep chocolate.

The only opaque glass produced by nineteenth century bottle-makers was 'milk glass'. It was made by adding tin or zinc oxide to the crucible. This produced white glass which could be used to make 'blue milk' or 'green milk' by the addition of different amounts of copper oxide. The opaque 'milk glass' bottles blown from it were used mainly by cosmetics manufacturers.

Unlike wine merchants and brewers, who did not wish their customers to see the contents of bottles too clearly, fruit bottlers and pickle makers were eager to let the public know that their containers held nothing but wholesome food. Throughout most of the nineteenth century they had to be content with pale aqua bottles because it was impossible until the 1890s to produce colourless glass at a price at which it could be used to make cheap bottles. But just before the end of the century it was found that manganese added to the crucible produced an inexpensive glass that was almost colourless. Its only disadvantage was that bottles made from it took on a purple tint if exposed to strong sunlight for long periods. This defect did not greatly trouble food packagers in northern Europe where the sunshine was rarely of sufficient intensity to effect the bottles; but in America, Australia, and other sunny climates precautions had to be taken when storing bottled fruits and sauces on shelves or displaying them in shop windows to ensure that the glass did not turn purple before the containers were sold. Supplies of the manganese used to make this colourless glass came mainly from Germany. When it became impossible to obtain the chemical during the war years of 1914–18 glassmakers in Britain and America were obliged to find an alternative. This proved to be selenium which produced colourless glass unaffected by sunlight. It has been used for this purpose ever since.

8 *Stoneware bottles*

For more than two centuries, from the first widespread use of commercial containers in the seventeenth century to the mechanization of the industry in the twentieth century, potteries fought a losing battle with glassworks in competition for bottlemaking contracts. Nevertheless, a substantial number of stoneware bottles were made, and in one or two specialized markets potteries won the bulk of the orders for many years. German wine exporters abandoned their use of salt-glazed Rhenish stoneware in the eighteenth century, but bottlers of German spa water remained loyal to their potteries until the sale of natural mineral waters ceased at the end of the nineteenth century. They used cylindrical salt-glazed bottles into which their trademarks were cut, or incised, with metal stamps. Examples have been found in late nineteenth century dumps throughout the world. In England and America vinegar and cider makers preferred stoneware bottles well into the twentieth century, as did bottlers of porter and stout in Ireland and parts of Scotland. The vinegar and cider bottles were often of large capacities and many were sold by the brewers to retailers who filled customers own bottles from them. The porter and stout bottles, which were sold over the counter, imitated the shapes of rival glass bottles; early specimens had wide, squat bodies, while later versions were slim and straight-sided.

Between 1820 and 1860 several English potteries produced a range of elaborately moulded spirit flasks, most of which, like the historical flasks made by American glassworks at about the same time, commemorated important events in British history or portrayed notable personalities of the period. Some of these flasks were figurals depicting Sir Robert Peel, Daniel O'Connell and other politicians, while members of the royal family including Victoria and William IV were also popular models. Many had political slogans or the name of the person portrayed incised in the clay. Some potters eschewed the three-dimensional figural form and simply moulded portraits of the models on the bodies of the bottles. A very few produced decorative but non-commemorative flasks in shapes including mermaids and merboys.

In all about one hundred different stoneware flasks were made. Judging by the numbers which have survived in museums and private collections, and which have been recovered by dump diggers, they proved very popular with hard liquor drinkers of the period. Why such a lucrative market was not exploited by bottlemakers is not known, but no historical flasks made in glass have yet been found in Britain. Perhaps the intricate moulding techniques necessary to produce these delightful bottles were beyond the capabilities of bottlemakers at that time. Bearing in mind that it was not until 1821 that Henry Ricketts patented his base-embossing three-piece mould it seems not unreasonable to surmise that glassmakers were unable to compete with potters in this market. Weight is given to this theory by the fact that potters continued to exploit the figural bottle market well into the third quarter of the nineteenth century. The items they made included hot water bottles and muff warmers in the shape of ladies' handbags, wreath-shaped bottles, and other convoluted designs which were possible when working with clay because each piece could be press-moulded in several different parts, the whole being stuck together before firing.

Although the bulk of the market for food packaging bottles was lost to glassmakers when the need arose for containers which allowed the customer to see the contents, potteries were still able to find markets for stoneware bottles for those household products which the customer did not wish to see before buying. Boot blacking, grate polish, ink for use in schools and offices, paint, varnish, and other oil-based products were all sold throughout the nineteenth century in brown, salt-glazed stoneware bottles. But by far the largest market which potteries were able to retain in spite of competition from glassworks was the supply of ginger beer bottles.

Notwithstanding the enormous popularity of artificial mineral waters ginger beer remained the most widely consumed of all soft drinks until well into the twentieth century, and apart from a few 'black glass' ginger beer bottles which appeared in the late nineteenth century mineral water makers used stone bottles for this drink for almost one hundred and fifty years. (Indeed it was popularly called '*stone* ginger beer'.) Until the 1850s the bottles were of a brown, salt-glazed colour with trademarks incised on their shoulders or bodies. When glassmakers began to emboss trademarks on bottles potteries competed by using transfer-printing techniques to mark ginger beer bottles with indelible labels ideally suited for use on bottles which were washed and re-filled many times during their useful lives. The transfers were usually black and printed against a white glaze background, but in the 1880s when coloured glass was widely used for mineral water bottles many potteries turned out ginger beer bottles with brightly coloured glazes. The best of these colour-printed specimens were made by Scottish potteries and their attractiveness encouraged many Scotch (and some Irish) whisky distillers to order similarly decorated stoneware bottles for their export blends. Many of these have turned up in dumps in Australia and the United States.

9 The men who made the bottles

Writers and commentators who recorded what few facts were set down about bottlemaking in the nineteenth century concentrated most of their attention on inventors and on those owners of glassworks who made fortunes from the industry. Of the craftsmen who worked at the furnaces and who blew and moulded the inventors' creations in order to make the owners' fortunes precious little is known. A search of more than a thousand newspapers and dozens of books on nineteenth century technology and working class life produced barely enough information to fill a couple of pages. Nevertheless, it is possible from the brief glimpses we are allowed to gain at least some idea of what life was like for the men and boys who made the bottles now treasured by present-day collectors.

In the 1830s boys were taken on at the age of eight. (It was not until the 1890s that the starting age was raised to twelve). They did not serve an apprenticeship as we understand the term today. There was no set training programme; a bright lad learned the craft by watching his father and by practising with the tools during meal breaks. If he showed an aptitude for the work he might rise in ten or more years to the position of blower; but many men spent their entire lives at the job without ever blowing bottles. In addition to acting as general labourer to all and sundry, the boy's main duty was to keep the bottle moulds at the correct temperature. If wooden moulds were employed he held a piece of hot glass on a pontil rod inside the mould after the blower had removed a blown bottle until the mould was required to form the next bottle. If metal moulds were used there was

18., 19. Bottlemakers at work in the nineteenth century.

usually a small subsidiary furnace at which the boy kept a dozen or more moulds in a state of readiness for several blowers. By moving them in and out of the furnace he maintained their temperature at just below red-heat. If the moulds were allowed to become too hot glass adhered to them during blowing; if too cold they imparted a 'ruffled' surface to the bottle. For either of these offences the boy could expect a cuff from one of the men. After three or four years of mould warming a 'likely lad' would take his place among the older men and learn to use the various empontilling and neck-forming tools.

At some time during the working week it was necessary for all hands – boys and men – to prepare batches of glass and crucibles for future work. Crucibles, or pots, were made by mixing fresh fire-clay with the crushed clay from old pots. Kneading of this clayey mass was done with the bare feet after which the pots were formed by hand and set aside to dry. A contemporary account of the operations that followed vividly describes the dangers and difficulties of the work.

'When the pots are dry, they may be removed to a hot room of one hundred to one hundred and fifty temperature. Before setting in the glass furnace great care is necessary to anneal a pot in the arch; a week or more should be allowed gradually to bring it to a white heat, ready for pot setting. This work is always performed towards the end of the week, and is a hot and fatiguing operation; all hands must be present, and absentees, except for illness, are severely fined. The men are provided with suitable dresses to shield them from the open blaze of the furnace. The old pot, being no longer useful, by age or accident, is then exposed, by pulling down the temporary brickwork; a large iron bar, steeled and sharpened at the point, is placed across another bar, to operate under the pot as a fulcrum; several men rest their entire weight upon the end of this long lever, and, after one or many efforts, and perhaps many more simultaneous blows of the bar, used as a sort of battering-ram, the old pot, either wholly or by pieces, is detached from the siege of the furnace.

Continued on page 129

1 Eighteenth century black glass wine bottles from Britain; two with seals, one unsealed.

2 Three bottles made in Bristol after 1821 in Ricketts' moulds.

3 Bases of bottles in Plate 2 showing the embossing usually found on Ricketts' moulded bottles.
4 Wine bottle with low body seal made in the 1840s.
5 Group of sealed bottles showing evolution of lips on British wine bottles from the eighteenth to the late nineteenth century.

A Dutch case gin bottle with shoulder seal made for A. van Hoboken, Rotterdam, in the 1880s.

7 Three interesting European sealed bottles recovered from a dump in Holland.

8 Three late nineteenth century British-made wine bottles.

9 A group of maraschino liqueur bottles from Zara, Dalmatia, now part of Yugoslavia.

10 European wine bottles made in the late nineteenth and early twentieth centuries.

An amber liqueur bottle from Greece.

12 Liqueur bottle made in Greece in the 1890s.

13 Two European sealed wine bottles in an unusual shade of olive-green glass.

14 Round-bodied wine and spirit bottles from America and Britain.

15 French liqueur bottle made in the early twentieth century.

16 A Spanish drinking bottle wi[th] pontil scar; probably made in t[he] 1850s.

17 German hock bottles in red glass made in the 1920s.

18 Embossed beer bottles from Britain with ornate trade marks typical of British beer bottles in the late nineteenth century.

19 Swing stoppered bottle with stopper secured and closed by wire 'bails'. Made in Britain in the 1890s.

20 A late nin teenth centu beer bottle fro Trieste.

21 Late eighteer century sto ware spirit b tle from E land with name and dress of wine mercha who used it cised in body.

2 Stoneware spirit flask made in England during the reign of William IV.

3 A figural flask in the shape of a man sitting on a barrel. Made in England in the 1830s.

24 A mermaid figural probably made by Doulton & Co. of Lambeth.

25 a, b, c. Historical whisky flasks made in the United States before 1860.

26 Front and rear views of a decanter used in a Victorian public house in the 1860s.

Two European stoneware bottles of the nineteenth century.

Two transfer-printed bottles made in England before 1900.

29 A colourful ginger beer bottle from Wales.

30 Transfer-printed Scotch whisky bottle made for the export market in the late nineteenth century. This specimen was found in Australia.

31 A whisky bottle made by Doulton & Co. of Lambeth with a transfer printed picture of an English ale house.

32 Embossed 'coffin flask' whisky bottle of a type used in Britain in the nineteenth century.

33 Glass whisky jug made in England in the 1890s.

34 a, b, c. Late nineteenth century American spirit bottles.

A colour-printed stoneware chemist's jar sold in Britain, America and Australia in the late nineteenth century.

Stoneware hot water bottles from Britain.

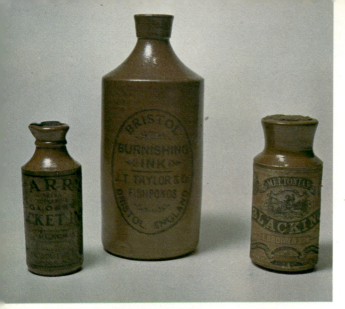

37 Two labelled blacking pots and a bulk ink bottle, used in Britain throughout the nineteenth century.

38 An ammonia b[ottle] in stoneware; m[ade] in Scotland aro[und] 1900.

39 Salt-glazed stoneware ju[g] made in the United State[s] in the 1870s.

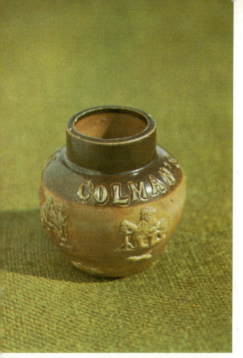 

A British mustard pot as sold in Britain in the 1860s.

41 A glass mustard pot from France.

42a, b. American sauce and pickle bottles in aqua glass.

American vinegar jug embossed with a picture of The White House. Early twentieth century.

44 American fruit jar with metal lid.

45 German stoneware bottle used in the nineteenth century for exported natural mineral water.

46 Three glass spa water bottles used in Europe in the nineteenth century.

47 A Hamilton bottle complete with original contents, cork and label.

48 An early globe-stoppered Codd bottle made before 1881 at the Stairfoot Bottleworks, Barnsley.

49 A flat-bottomed Codd-Hamilton hyb

50 A rival to the Codd bottle; this is a Deek patent of 1885.

Another globe-stoppered mineral water bottle; an early twentieth century Haines' patent.

52 An 'anti-theft' bottle with red lip made by Dan Rylands in the 1880s. The coloured lip made the bottle easy to identify.

53 A blue-lipped Rylands' bottle. This is a 'Bulb' Codd.

54 A dark green Codd made by one of Dan Rylands' competitors in the 1880s.

55 A rare cobalt-blue Hamilton bottle used by a mineral water maker from Newcastle-upon-Tyne in the 1890s.

56 An olive-green Hamilton.

57 A rare 'improved' Hamilton in blue glass made in Britain in the 1890s.

Codd bottles with coloured marbles.

59 'Bullet-stoppered' bottles made in Britain in the late nineteenth century.

60 Embossed internal screw-stoppered bottle of the type which replaced globe and bullet-stoppered bottles in the twentieth century.

61 A rare screw-stoppered bottle with a valve fitted to its ebonite stopper. The valve was depressed to release gas pressure before opening the bottle.

62 'Swing-stoppered' bottles used by a few mineral water makers in Britain in the 1920s.

63 All the bottles shown here were used by a British mineral water maker in the nineteenth century.

64 a, b Hutchinson soda water bottles used in the United States in the late nineteenth century.

65 Two early chemist's shop display bottles from England.

66 A group of early Victorian medicine bottles from Britain.

67 Daffy's Elixir bottles. The medicine was popular in Britain and America in the early nineteenth century.

68 A three-sided patent medicine bottle from Britain.

69 This wedge-shaped patent medicine bottle once held cough mixture.

70a, b Two late nineteenth century American patent medicine bottles.

71 Bottles used in the nineteenth century as containers for the patent medicines of H. H. Warner, a famous American 'quack' doctor.

72 A miniature Warner's bottle shown alongside a half-pint specimen.

Bottles used for William Radam's 'Microbe Killer', another famous nineteenth century medicine sold in Britain and America.

74a, b, c   American bitters bottles.

75 Hair restorer bottles sold in Britain, North America, and the British colonies in the nineteenth century.

76 Two rare babies' feeding bottles: a boat-shaped feeder of the early eighteenth century, and a purple glass specimen probably made in America in the late nineteenth century.

77 A group of British baby feeders made in the late nineteenth and early twentieth centuries.

78 A group of British poison bottles made between 1880 and 1910.

79 Early twentieth century British poison bottle.

80 Two figural
a, b poison bottles from the United States.

81 A group of perfume bottles from Britain.

82 Milk-glass bottles from Britain.

83 A milk-glass bottle found in Australia.

84 Pocket spitoons sold in Britain in the nineteenth century.

85 European poison, cologne and perfume bottles made in the late nineteenth century.

European Bottles
(85 continued)

European Bottles
(85 continued)

European Bottles
(85 continued)

86 A fine display on the shelves of a collector who specialises in chemists' bottles.

87 Another view of the same collection showing labelled bottles used by British chemists in the late nineteenth century.

88 Five rare ink bottles from Britain, all made in the nineteenth century.

89a, b Stoneware and glass figural inks from the United States.

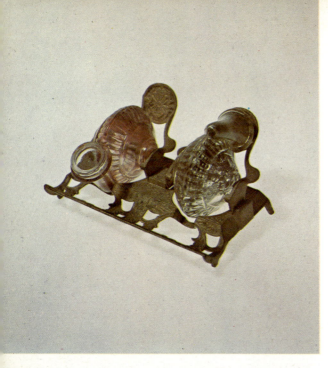

90 A metal inkstand with original bottles probably made in America in the 1870s.

91 American bulk ink bottle in blue glass.

92 A bulk ink bottle in aqua glass made in Greece in 1900.

93 A French ink bottle of the 1890s.

94 Glass fairy lights once used to decorate Victorian Christmas trees.

95 Oil lamps from British and Australian dumps.

96 Glass fire grenades filled with carbon tetrachloride. They were thrown into the centre of a fire where they shattered and smothered the flames with their chemical contents.

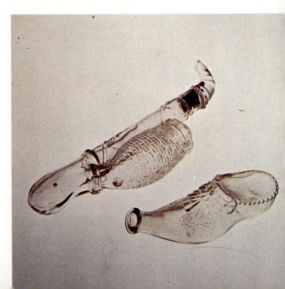

97 Three unusual figural bottles from Poland.

98 A Spanish figural bottle probably made in the 1860s.

99 A bear figural from Germany in deep purple glass.

100 A swan figural made in Greece in the 1920s.

101 A bow-shaped perfume bottle made in Britain in the 1920s.

102a, b   Avon cosmetics bottles

102c, d, e Hair dressing bottles used in barbers' shops.

102f Candy jar

102g, h, i, j, k, l, m Miniature spirit bottles

103 A colourful display of antique bottles from Britain, America, Europe and Australia.

About six or eight men take afterwards each a bar about five feet long, like a javelin steeled and sharpened at one end; they rush forward in face of the fiery furnace, guarding their faces with their protected arms, and aim a blow at such of the irregular rocky incrustations of clay as adhere to the siege. This operation is repeated until the pieces of partially vitrified clay are wholly removed from the position on which the old pot stood, which should be repaired with clay and sand. The new pot, at a white heat, is then removed from the annealing pot-arch, and carried upon the end of a two-wheeled iron carriage with a long handle, by four or more workmen, who carefully set it or tilt it backwards into its proper position in the furnace.

In the interim between removing the old and setting the new pot, an iron screen is placed before the opening of the furnace, which, having lost most of its heat, is urged gradually to its original high temperature. Seldom more than two pots are set in one week; to do more would endanger the other pots in the furnace, by their getting too cold, through the furnace remaining so long open. The fatigue and exhaustion of the men, who are often detained four hours in this operation, is also very great, and is attended occasionally by severe falls, burns, or bruises, by liability to catch cold, great excitement, energetic exertion, and exposure to the flame of the open furnace.'

In Britain at that time there was no compensation for injuries sustained at work. Invariably managers and owners claimed that burns, falls, and other mishaps were either the unavoidable and uncompensated risks of the job, or were the results of 'larking' by the men and boys. One accident described by a contemporary writer gives some idea of shop-floor dangers.

'A man was severely scalded, and confined to his house about six weeks, through falling back after pot setting into a caldron filled with water, standing in the glasshouse, into which the remaining contents of the pots had been emptied at the end of the week; this did not occur in the course of his duties, but by what is termed 'larking'. He was dreadfully scalded, but not burnt with the fused glass. A pump being in the glasshouse, cold water was abundantly poured upon the scalded parts, the man was carried home, and attended by a skilful medical man; after several weeks of suffering, he recovered, and returned to his work.'

American glassworkers fared rather better at the hands of their masters. Dr. Dyott, probably imitating other American glassworks owners, certainly compensated workmen injured during the course of their duties, as well as providing free schooling for young boys, decent housing for glassworkers' families, and other benefits almost unheard of in Britain in the nineteenth century. It is hardly surprising that many British glassworkers were prepared to risk the passage to the New World whenever the opportunity presented itself, as it so often did when a company's order books were half empty and many workers laid off without pay.

Little is known about the numbers of bottles made during a working day or about wages paid to employees. The only figures found during my researches refer to the production of small apothecaries' vials made in two-piece moulds. A group of three blowers and two boys, called a 'chair' in the terminology of the industry, were expected to produce two hundred vials in the course of a six-hour shift for which they received (in the 1830s) five shillings between them. In practice an experienced 'chair' could double that figure and thereby earn ten shillings per shift which they probably divided as three shillings each for the men and sixpence each for the boys. The average earnings quoted for men was twenty-seven shillings per week. In order to earn that amount a 'chair' had to work two six-hour shifts (or 'moves') in the course of a twenty-four-hour day. Thus a man would report for duty at six a.m. and work until two p.m. He then went home, reporting back for his next shift at ten p.m., working twelve 'moves' in a six-day week. Because of the unusual hours many glassworkers kept public-houses as a side-line to augment their incomes. Wives and daughters assisted in running the pub in order to keep open for as long as possible. No doubt many of these men made, during slack periods at the works, bottles and decanters for use in their own drinking establishment.

Key to colour plates

Plate 1 – Three eighteenth century 'black' glass wine bottles from Britain. The two on the left have body seals bearing the initials W.A.; the specimen on the right is unsealed. All three bottles have crude pontil scars and deep kick-ups. Prior to the invention of Henry Ricketts' mould in 1821 it was almost impossible to form bottles of uniform shape and capacity. Cylindrical sides were formed in one-piece cup moulds, but when the bottle was removed from the mould and the pontil rod applied to hold the bottle before forming the lip (and later to form the kick-up) this caused the base of the bottle to expand outwards and produced the 'bell' shape as seen on the lower bodies of these examples. Note the drunken stance of the bottle in the centre; this was caused by the pontil rod being applied to the base in an off-centre position. Although their lack of uniformity and crudity of shape make these early wine bottles highly desirable to present-day collectors, they were quite unsuited to the needs of nineteenth century bottlers and merchants who required containers of symmetrical shapes and standard capacities, especially when machines were invented that could fill, cork, and label bottles at great speed.

Plate 2 – Three Bristol bottles made after 1821 in Ricketts' moulds. Note the uniformity of shape and capacity and the absence of bell-shaped bases. The three-piece mould seams can also be clearly seen on the bottle on the right, and the word PATENT has been embossed on the shoulders of the bottle in the centre. As on many Bristol bottles the seals are positioned centrally on the body and the stringrings take the form of flaired skirts beneath the lip. Ricketts' bottles achieved international fame and were bought by wine merchants and beer bottlers throughout the world. His success resulted partly from Bristol's proximity to cheap coal from South Wales and partly from its trade links with America and the British colonies. Shiploads of Ricketts' bottles were despatched to the United States, the West Indies, and Australia in the nineteenth century, though the United States market was lost after 1850 to American bottlemakers. Some of the Bristol bottles found by diggers in the southern states of America and near Botany Bay in New South Wales must surely have found their way there as cargo on slave and convict ships.

Plate 3 – Bases of three Bristol bottles in Plate 2 shown alongside a bottle not made in a Ricketts' mould. Two the of lower specimens carry the words H.RICKETTS.GLASSWORKS.BRISTOL; the upper bottle has the words POWELL.BRISTOL embossed on its base. Powell & Filer amalgamated with Ricketts' Glassworks in 1857 and for several years afterwards the company's bottles were embossed POWELL & RICKETTS; but Ricketts' name is not seen on bottles made by the company after 1860.

Ricketts' patent protection on his invention lapsed in 1835; thereafter any bottlemaker could copy his mould without infringement of patent rights. That they did so is proved by the numerous bottles recovered by diggers which do not carry the name of a Bristol company but which are identical in every other way to bottles made in that city. Many have turned up in sites around Leeds in Yorkshire and in the Tyne and Wear areas of County Durham, both regions where the making of bottles became important industries in the late nineteenth century.

Plate 4 – In the 1840s and again in the 1880s a small number of wine merchants ordered bottles with seals positioned low on the body so that a label could be fixed to the bottle on the same side as the seal. When seals were placed centrally on the body the label had to be attached on the other side. Most wine merchants overcame the problem by ordering bottles with seals placed on the sloping shoulders. This specimen was probably made in the 1840s.

In 1974 the British Bottle Collectors Club ran a competition to find the bottle with the lowest body seal owned at that time by a British collector. The winning bottle was a black glass beer bottle with the outer edge of its seal just half an inch from the base of the bottle. During the competition many other specimens, wines and beers, with seals not more than one inch from their bases were exhibited by collectors.

Plate 5 – This group of five sealed bottles shows, from left to right, the evolution of lips on British wine bottles from the eighteenth to the late nineteenth century. When attempting to date a bottle it is essential to consider other factors in addition to lip shape because all glassmakers did not progress at the same rate and many continued to use old-fashioned lip-forming tools throughout the nineteenth century. Additionally wine merchants often ordered bottles of old-fashioned shape to be used when bottling very old wines. Collectors must also look at body shape, base, and seal before estimating the age of a bottle. Even then they could err by as much as thirty years in their dating. The only reasonably certain method of gauging a bottle's *latest* possible year of manufacture is to excavate it from a refuse dump of precisely known age. This can often be ascertained by checking official records held by town and city archives where the minutes of corporation refuse disposal committees are kept (in Britain) for as far back as 1850. Once the year in which a refuse dump was finally abandoned has been ascertained one can state with confidence that bottles and other objects excavated from it must at least date from the last year of the dump's use.

Plate 6 – A Dutch case gin bottle carrying a shoulder seal impressed with the letters AVH and embossed on its body with the name of the distiller, A van Hoboken Rotterdam. Square case bottles were first used in the seventeenth century to meet the need for bottles which could be packed safely in square wooden crates. Filled with Dutch gin, they were exported to Britain and France (and later to America and Australia) to satisfy demand for this world-renowned spirit. Unsealed bottles of the same shape were used up to the 1920s, and some present-day gin distillers still favour square bottles.

Many of the Dutch gin and schnapps makers who controlled the export trade in the late nineteenth century centred their operations at Schiedam where vast distilleries producing millions of gallons of spirit were built. Each company used olive-green square bottles identified either by shoulder seals bearing trademarks and brand names or by embossing on the body. A paper label was also applied to the side of each bottle. By 1900 the use of seals and embossing had almost completely disappeared, though a few companies continued to emboss the word SCHIEDAM on one side of the bottle.

Plate 7 – Three choice sealed bottles recovered from a dump in Holland. *Left*: A Dutch gin bottle of unusual shape used by P. Loopuyt & Co., Schiedam. The seal indicates the bottle once contained 'Hourglass Gin', a brand widely advertised in the 1880s. Bottles bearing similar seals but of the more conventional case bottle shape have been found in Britain and Australia. *Centre*: A maraschino liqueur bottle used by F. Drioli, Zara, Dalmatia. *Right*: A German bottle thought to have contained bitters. The seal, a most unusual embellishment on a bitters bottle, carries a star and fish, trademark of Hartwig Kantarowicz, Posen, Germany, bitters manufacturers who sold their products throughout Europe and North America in the late nineteenth century. As with most bitters bottles, this specimen is also embossed with the company's name.

The dump from which the bottles were recovered was in an area of former marshland reclaimed in the 1890s. The finds prove beyond doubt that rich rewards await those who pursue the hobby in Holland where vast amounts of land were reclaimed in the nineteenth century by the dumping of household refuse, and where at the present time there are only a handful of excavators at work.

Plate 8 – Three sealed wine bottles made in the late nineteenth century. By the 1860s most glassworks were making bottles of uniform shape and capacity, though a few small companies continued to make bottles by the old-fashioned free-blown methods. To add to the problems of present-day collectors who attempt to date specimens in their collections many seals were impressed with vintage years which can be mistaken for a bottle's date of manufacture. The specimen in the centre of the photograph has a seal bearing the words ROUSDON JUBILEE 1887, an obvious reference to the bottle's contents and not to its year of manufacture which *might* have been as late as 1900.

In addition to the use of coronation and anniversary years on their seals, wine merchants were also fascinated by those years in which comets were observed in the skies. Indeed, the term 'comet wine' is used to this day in the wine trade to indicate a wine of superior quality. For many years after the appearance of the Great Comet in 1865 thousands of wine bottles were made with that date on their seals.

Plate 9 – A group of maraschino liqueur bottles from Zara, Dalmatia – now part of Yugoslavia. In the eighteenth and nineteenth centuries Dalmatia was part of the Austro-Hungarian Empire and the liqueur makers of Zara incorporated the spread-eagle emblem of the emperor in the trademark used on their bottles. Throughout the eighteenth and for most of the nineteenth century the bottles carried glass seals, but in the late nineteenth century seals gave way to paper labels and an embossed 'seal' as seen on the bottle on the extreme right. The spread-eagle emblem was not used on the embossed bottles because by that time the country was no longer a province of the empire.

Dump diggers throughout the world have found examples of these beautiful bottles which were made in delightful shades of green and turquoise glass. Early specimens, like the one on the left, had sheared lips; applied lips were used after 1860. Round Zara bottles have also been found. Although all the seals carry the spread-eagle and the word ZARA, they have different company names around their borders. Most common are those bearing the name, Drioli, while bottles bearing the names Stampali, Maggazin and Luxardo have been found in smaller numbers. The Luxardo company is still in business.

Plate 10 – A group of late nineteenth and early twentieth century wine bottles from Europe. Note the collar-below-lip seen on the green specimens. This type of lip finish and the lighter green glass are typical of European wine bottles of this period. The deep kick-ups on the green specimens were probably a method of strengthening the bases of the bottles; they were certainly not made by pontil rods because all of these bottles were machine-made and it was unnecessary to work the lips by hand. The amber specimen, thought to date from the 1920s, has a plain seal which probably contained a small paper label. This type of seal, often embossed rather than applied, is still occasionally seen on modern European wine bottles.

Plate 11 – An unusual long-necked amber bottle, possibly a liqueur, but origin unknown. The base is embossed PIXAVON. It was bought at a flea-market in Athens where the sale of old bottles for re-use as containers for home-made wines and olive oil is common. Many nineteenth century bottles have been re-sold and re-used many times and still find their way back to the flea-markets whenever a house is cleared of unwanted possessions. By visiting these markets when on holiday in Greece many British and American collectors have been able to buy very desirable additions to their displays for small sums of money. Similar bargains can be found in other Mediterranean countries.

Plate 12 – A round liqueur bottle used in Greece in the 1890s. Liqueur and spirit bottles of this shape enjoyed a period of worldwide popularity at the end of the nineteenth century, though the use of aqua glass appears to have been confined to continental Europe. British examples were usually made in black, brown, or dark green glass. The 'collar-below-sheared-lip' finish seen on the neck of this specimen is not found on similar bottles from Britain, but it was used on some bottles made in the United States where continental European influence was much stronger than it was in Britain thanks to massive emigration from Greece, Italy, Germany, and Eastern Europe. The emigrants took with them to the New World their tastes for European food and drink which they must have preferred to buy in containers similar to those used in 'the old country'.

Plate 13 – Two late nineteenth century European sealed wines made in an unusual shade of olive-green glass. Glass of this colour was commonly made by adding oxides of iron and copper to the crucible, but in Holland where sea sand containing substantial amounts of iron oxide was used in glassmaking it was only necessary to add copper oxide to the crucible to produce the olive-green glass used for most Dutch gin bottles. Glass of a similar shade was produced in England in the late nineteenth century at glassworks on the northeast coast where beer bottles were manufactured. As in Holland, sea sand was used in the process. Copper ores are found in small quantities in the high moorland areas of this part of Britain; if they washed down to the coast in sufficient quantities it is possible that olive-green glass was produced without the addition of more copper oxide.

Plate 14 – Round-bodied wine and spirit bottles from America and Britain. The two on the left are American; the British specimen on the right held 'BIG TREE BRAND' wine (thought to have been a colonial wine from Australia) and employed an internal screw stopper. All three bottles date from the late nineteenth century.

Consumption of colonial wines was high in Britain before 1900. Vines were successfully introduced into Australia and South Africa in the early nineteenth century and sales of wines from these countries rose steadily thanks to low import duties. Much of the produce was bottled in England and re-exported to the colonies. It was also used to make British brandy.

Plate 15 – Early twentieth century French liqueur bottle embossed DUMONT. The misaligned neck is reminiscent of early free-blown bottles, but this example was mould-blown; the neck probably sagged as the bottle cooled after removal from the mould. The embossed circle on the shoulder is often found on bottles used by companies which applied seals to their bottles in earlier times. In most cases this circle held a small paper label printed with the same design used on the original seal. A few wine merchants filled the circle with sealing wax and impressed a seal into it after filling and corking the bottle. Sometimes the string or wire used to secure the cork on bottles containing effervescent wine was placed behind the wax seal. When the bottle was opened the flying cork pulled out the wire and broke the seal, thus preventing illegal re-filling of the bottle.

Plate 16 – Spanish drinking bottle used for wine. Similar bottles are still made in Spain and Portugal, but this specimen has a pontil scar and is thought to date from the 1850s. It was recovered from a Madrid refuse dump by an enterprising British collector who took his dump-digging tools on a Spanish holiday. Collectors from the

United States have found identical specimens when digging dumps in South America.

Plate 17 – German hock bottles in red glass made in the 1920s. Similar bottles are still in use in the German wine trade today. Although bottle collecting has not yet achieved widespread popularity in Germany many German bottles dating from the nineteenth century have been recovered by American armed forces personnel who have pursued the hobby while on tours of duty in that country.

Plate 18 – In Britain beer bottles followed a pattern of evolution similar to wine bottles, though the use of seals was abandoned much earlier. By the late nineteenth century almost all British beer bottles were richly embossed with ornate trademarks as seen on these two specimens. Embossing was favoured by bottlers who re-used empties because paper labels invariably fell off during washing. Dark green glass was most popular for beer bottles, but brown, black, and olive-green bottles were also used. Most beer bottlers used corks until the 1890s when internal screw stoppers came into widespread use. The earliest of these screw stoppers were made from wood, an air-tight closure being achieved by the use of a rubber washer above the threads on the stopper. Later stoppers made from ebonite came into fashion. These were usually incised with the company's name and a request to the customer to 'replace the stopper before returning the bottle'. Internal screw stoppers were in turn gradually replaced by 'crown' caps made from metal in the twentieth century.

Plate 19 – Swing stoppers had the great advantage that they could not be lost by customers, but their disadvantage was that the wire 'bails' which fixed the stopper securely to the bottle rusted very quickly. For this reason they were never very popular with beer bottlers. This half-pint specimen dates from the 1890s. Note the word IMPERIAL embossed on the shoulders. This certified that the mould in which the bottles were made had been passed as accurate by Weights and Measures inspectors. The word is also seen on many stoneware jugs which were used in Victorian pubs to measure pints and half-pints of draught beer before the widespread introduction of beer engines.

Plate 20 – An 1890s beer bottle from Trieste embossed TRIESTER

EXPORT BIER BRAUEREI. It is quite unlike British beer bottles of the period and has a collar of glass beneath the lip and a kick-up in its base. American and Australian beer bottles of the late nineteenth century have similar lips known as 'ring seals' by collectors in those countries. Lager-type beers, which achieved great popularity in Australia and America in the nineteenth century, were exported from Europe in bottles like this until the 1930s. When Australia and America began brewing and bottling their own beers they used similar bottles.

Plate 21 – A stoneware spirit bottle from England, typical of the wares produced by potteries at the end of the eighteenth century when glass bottles began to capture the bulk of the market. By 1850 most of the sales for wine, beer, and spirit bottles had been won by glassmakers; only ginger beer bottlers and a few whisky distillers continued to use stoneware. Note the incised name and address of the merchant, Summerby of Grantham. This method of marking stoneware bottles was later superseded by the use of transfer-printing, but during the first half of the century it provided an inexpensive means of marking stoneware bottles that was far superior to paper labels. Some companies had pictorial trademarks incised on the bodies of their stone bottles and company names cut into the shoulders. A few imitated glass sealed bottles and had pads of clay placed on the shoulders into which a seal was impressed.

Plate 22 – A stoneware spirit flask made in England in the reign of William IV. The scroll carried by Mr Punch bears the words THE TRIUMPH OF THE PEN and the flask is thought to commemorate the passing of Catholic Reform laws in the 1830s. It is one of a number of flasks bearing political propaganda produced in Britain in the early nineteenth century. They are thought to have been made for and sold by innkeepers who supported the Catholic Reform movement. Several famous potteries including Bourne of Denby and Doulton of Lambeth are known to have made these bottles.

This specimen was recovered from a dump at the rear of a large Victorian mansion. The present owner decided to extend the lawn at the back of the house and while rooting up a large nettlebed he came upon a vast accumulation of early nineteenth century throw-aways which included several stoneware flasks. The refuse had been dumped less than one hundred yards from the old kitchens and the lucky finder was able to obtain a valuable collection of bottles without venturing beyond his own back garden.

Plate 23 – A figural stoneware flask in the shape of a man sitting on a barrel enjoying his favourite drink. Although this example is unmarked other versions of the bottle have the words REFORM CORDIAL incised on the lower part of the body, an indication that they were also produced as political propaganda in the 1830s. The brown, 'treacle' glaze is found on many of these bottles. This particular specimen is of half-pint capacity, but larger versions, including half-gallon bottles probably used in public houses, were also made. The larger specimens have survived in greater numbers than the small pocket flasks.

Other stoneware flasks bearing political slogans or made in the image of political leaders of the early nineteenth century include: Daniel O'Connell figurals bearing the words IRISH REFORM CORDIAL; William IV figurals bearing the words WILLIAM IVth REFORM CORDIAL; Lord Brougham figurals bearing the words THE TRUE SPIRIT OF REFORM and THE SECOND MAGNA CARTA; Lord Grey figurals bearing the words GREY'S REFORM CORDIAL; Lord Russell figurals bearing the words THE PEOPLE'S RIGHTS and THE TRUE SPIRIT OF REFORM; Sir Robert Peel figurals bearing the words REFORM AND REPEAL and BREAD FOR THE MILLIONS; and Cobden figurals bearing the words LEAGUE SUBSCRIPTION £200,000. R.COBDEN,MP. Another flask produced in the 1850s to commemorate the ending of the Crimean War shows three soldiers shaking hands and bears the words WAR DECLARED MARCH 28, 1854. PEACE PROCLAIMED APRIL 29, 1856.

Plate 24 – Not all stoneware spirit flasks made in Britain in the early nineteenth century were of a political nature. This mermaid figural is one of a pair made by Doulton (and others) between 1820 and 1850. Its companion was a merboy with beard and a less shapely chest! Like many more of these attractive stoneware bottles it was kept long after the contents had been drunk; a lucky digger recovered it on a late 1920s refuse site. A few weeks later another digger recovered a merboy from the same site. The pair were probably thrown out as unwanted junk during spring cleaning operations.

Queen Victoria, Prince Albert, and other members of the royal family were also popular models for makers of stoneware bottles. When the polka was introduced into Britain from France in the 1840s a few potters produced flasks in the shape of polka dancers. One or two notable entertainers including Jenny Lind were also depicted.

Plate 25 – The manufacture of whisky bottles provided early nineteenth century American glassworks with substantial orders and enabled them to compete with European glassworks. In order to promote sales whisky distillers ordered bottles moulded with eye-catching and often patriotic embossing. Shown here are an amber flask bearing the words 'FOR PIKE'S PEAK'; an aqua flask showing clasped hands to represent the Union; and an olive-green Cornucopia flask. All three were made before 1860.

Plate 26 – Front and rear views of a beautifully decorated flask thought to have been used as a decanter in a Victorian public house in the 1860s. The scenes depicted on the body of the bottle are typical public house activities of the period including snuff taking and clay tobacco pipe smoking. The use of different coloured glazes on the shoulders and necks of stoneware bottles became very popular after 1860. The crudity with which the brown glaze has been applied to this bottle suggests it was one of a large batch sold cheaply to many publicans rather than an expensive bottle made for a special customer.

Stoneware jugs and mugs decorated in a similar way and usually marked with the word IMPERIAL were also used to serve beer in Victorian public houses. Two of these in my own collection show hop-picking scenes. Like stoneware bottles these jugs and mugs gave way to glass during the second half of the nineteenth century, though their use alongside pewter mugs survived in country pubs into the twentieth century.

Plate 27 – Two European stoneware bottles of the nineteenth century. The wreath-shaped specimen is Polish; the boot is thought to be of German origin. As with many bottles from continental Europe now in British and American collections these two specimens were purchased in antique shops and are not thought to have been recovered from refuse dumps. Many more will no doubt be brought to light when dump digging achieves greater popularity in Europe.

Plate 28 – Transfer-printing was exploited most successfully by those potteries making ginger beer bottles. They were able to offer mineral water makers indelible labels in a wide choice of colours and guaranteed not to fall off when the bottles were washed. Several thousand transfer-printed ginger beer bottles were made in Britain between 1850 and 1920. Most prized by present-day collectors are those which carried pictorial transfers such as 'The Doctor's Stout' bottle seen on the right in this photograph. The beverage was nothing more than ginger beer flavoured with hops. The 'Double Stout' once contained in the bottle on the left was also a non-alcoholic beverage with a ginger beer base. Note the stone stopper used on the bottle; earlier ginger beer bottles used cork stoppers. Both of these bottles were made in Scotland by Buchan Potteries, Portobello, Edinburgh.

Stoneware bottles were ideally suited for ginger beer. The drink, which is a rather unattractive grey-brown colour, contains much yeasty sediment that settles on the bottom of the bottle during storage. By using stoneware bottles ginger beer makers were able to prevent this being seen by prospective customers. This was also the reason why the few glass bottle makers who attempted to compete in this market made their ginger beer bottles in very thick glass of the densest shades of dark brown and dark green.

Plate 29 – A blue-topped ginger beer from Wales. Other colours found on the shoulders of many late nineteenth century ginger beer bottles include red, green, and yellow. The best of the coloured specimens have been found by diggers in Wales and Scotland where ginger beer drinking was even more popular than in England and where fierce competition (there were often a dozen makers in a very small town) encouraged manufacturers to attract custom with eye-catching bottles. In Australia, where the drink was also well known, most of the bottles had black tranfers printed against yellow or white backgrounds. Until the 1870s almost all of these Australian bottles were made in England and exported empty to the Australian market. Colonial mineral water makers were able to order special transfers including their names and trademarks if they bought a minimum of ten dozen bottles. Most of the British-made ginger beer bottles shipped to the colonies were from the Lambeth pottery of Doulton & Co. The company's name is usually to be found stamped on the lower body of the bottle.

Plate 30 – Many whisky distillers were attracted to stoneware bottles in the late nineteenth century because transfer-printing offered a solution to the problem of paper labels which often fell off when bottles became damp during long sea voyages to overseas markets. It is for this reason that many Scotch and Irish transfer-printed whisky bottles are found in American and Australian dumps. This half-pint specimen, found in New South Wales, was made by Grosvenor Potteries, Glasgow, for the James Stewart Distillery. It contained 'Auld Lang Syne Brand' whisky.

During the writing of my, *International Bottle Collectors' Guide*, I visited Australia and saw at first hand the beautiful transfer-printed stoneware whisky bottles in many Australian collections. Although a few of these had been bought from diggers in Britain the majority were found in Australian dumps. When I returned to Britain I made a special effort to find and excavate sites in Scotland in the hope that some nineteenth century whisky distillers had used similar bottles for the home market. By a quite extraordinary coincidence one of the first specimens I found was identical to the bottle in the photograph; but very few other stoneware whisky bottles have yet come to light in the Scottish dumps so far explored.

Plate 31 – Not all orders for transfer-printed whisky bottles went to potteries in Scotland. This specimen was made by Doulton & Co. of Lambeth. It bears a transfer-printed picture of 'The Old Cheshire Cheese', a famous London alehouse. It is one of a series of Doulton bottles showing public houses and inns. Bottles bearing pictures of Scottish castles, Highland dancers, and kilted pipers were also made. Water jugs bearing similar transfer-printed designs have also been found.

Plate 32 – Most Scotch and Irish whisky bottled for the home market was packaged in glass bottles. The pint and half-pint sizes were either of round 'pumpkin seed' shape or, like the richly embossed specimen shown here, of sloping-sided 'coffin flask' shape. It was common practice in Victorian public houses to keep a stock of empty spirit bottles on the premises and to fill them for customers who wanted to take drinks home. They were sold in the 'bottle and jug bar', jugs being filled with ale. To identify his bottles the publican had the name of his establishment acid-etched on the body of the bottle. This service was provided by bottlemakers for those customers who could not afford to buy embossed bottles.

Plate 33 – A half-gallon whisky jug used in Britain in the 1890s. This particular specimen was made by Dan Rylands of Barnsley, maker of most of Britain's nineteenth century marble-stoppered mineral water bottles. Like all glass bottle makers of the period Rylands manufactured wares for many markets. His catalogue for 1890 includes five patented marble-stoppered bottles, a dozen different wine and spirit flasks, and a wide range of household bottles and jars.

Plate 34 – Many late nineteenth century American whisky bottles were as attractive as the earlier historical flasks. This photograph shows an amber bottle used for Booz 'Cabin' Whisky and an amber figural thought to have once contained gin. The aqua bottle is from the 'Kelly & Kerr Saloon', once patronized by Wild Bill Hickock and Calamity Jane.

Plate 35 – When colour printing became possible on stoneware many chemists ordered pots and jars with colour-printed labels. Shown here is a Confection of Senna jar issued by Boots Cash Chemists in the 1880s. The senna plant is illustrated in a green and yellow transfer on the body of the jar. Offered in three different sizes, the medicine was sold widely in Britain, Australia, and the United States.

Plate 36 – Three British stoneware hot water bottles made in the late nineteenth or early twentieth century. The purse-shaped figurals were both manufactured at the Bourne Potteries in Denby. Bed and muff warmers of this shape were very popular in Britain in the 1880s, small versions being carried by many ladies when travelling. The Govancroft Footwarmer was made in Scotland and is of a shape widely used throughout Britain until the 1920s; but the blue-glazed shoulders suggest that this particular bottle was made before 1900 because after that year most of these bottles were glazed in a single colour only. All stoneware hot water bottles were made in exceptionally thick pottery in order to retain the heat of the water for as long as possible. Rubber hot water bottles, familiar until the recent introduction of electric blankets, first appeared in the 1890s, but many people kept their old-fashioned stoneware bottles well into the twentieth century.

Plate 37 – Potteries were able to compete successfully with glassworks in supplying bottles for common household products. Shown here are two labelled blacking pots and a transfer-printed bulk ink bottle. They were used in Britain throughout the nineteenth century and thousands of transfer-printed specimens were despatched to colonial markets, this method of marking being preferred because paper labels fell off during transportation in damp ship-holds. Recent finds made during dump excavations in Scotland suggest that some English manufacturers regarded Scotland as an 'export market'; far more transfer-printed household bottles and jars have been found in Scottish dumps than have been recovered from sites south of the border.

Plate 38 – Another household bottle made in stoneware and transfer-printed in black against a white glazed background. The Plynine Company used stone bottles for its ammonia until the 1920s. Many of the bottles were made by the Buchan Pottery, Portobello, Edinburgh, renowned for the excellence of its ginger beer bottles. Note that the lower body of this bottle has been left unglazed and the surface of the clay roughened to give a better grip when the bottle is held in the hand. This safety measure reduced the risk of the bottle being dropped and its poisonous contents accidentally spilled. Some glass bottle makers copied this idea in later years by producing bottles for bleach and other corrosive liquids with deep flutings in their bodies.

Plate 39 – America's nineteenth century stoneware bottles were rather crudely printed and colour printing on stoneware was almost unknown. Shown here is a salt-glazed jug used by J. J. Duffy of The Eagle Tavern, Troy, N.Y. The tavern was destroyed by fire in 1876.

Plate 40 – An attractive mustard pot used by the British firm, Coleman & Co. in the 1860s. The use of raised decoration on stoneware bottles and jars was uncommon. Most were incised before 1860 and later types were more usually transfer-printed. This company is still in business; during two centuries of selling mustard they have used stoneware, glass, tin, and cardboard containers.

Plate 41 – A pot-bellied mustard jar from France. Glass mustard containers were not used in Britain where all mustard sold in the nineteenth century, including imported French mustard packaged by French manufacturers, came in heavy stoneware pots. The thickness of glass used for this jar suggests its designer copied a stoneware specimen, perhaps to make the container appear of greater capacity than it really was – a common practice in the nineteenth century. The use of one type of container for the home market and a different type for export sales has been noted in several countries and in a number of industries.

Plates 42/43/44 – Nineteenth century American household bottles for sauce, vinegar, pickles, and home-made preserves. These bottles were generally more attractively moulded than similar bottles used in Britain. The making of home-made pickles and preserves was enormously popular in rural America before 1900.

Plate 45 – A salt-glazed stoneware bottle used by the owners of a German natural mineral water spring. It is incised GEORGE KREUZEBERG AHRWEILLER EIN PREUSSEN. These bottles were made with and without handles and used throughout the nineteenth century for exported German spa water. On many specimens the finger prints of the potter can be clearly seen beneath the glaze, proof that the bottles were hand-made and not, like those used more recently by Dutch gin distillers, made by machine. Examples have been recovered from dumps in many parts of the world.

Plate 46 – A few German spa water bottlers used glass bottles as

seen on the left in this photograph. The other two specimens were used for natural mineral waters bottled in Britain. The centre bottle is embossed BOTTLED AT THE ROYAL PUMP ROOMS, HARROGATE. In spite of keen competition from artificial mineral water makers there was still a good market for spa water in Britain until the end of Queen Victoria's reign. In their advertisements bottlers stressed the medicinal qualities of naturally mineralized waters and claimed they would cure 'gout, liver complaints, and all disorders of the stomach'.

Plate 47 – Until the late 1860s, pointed-bottomed, Hamilton bottles were used for all mineral waters bottled in Britain, but after that date their use was limited to soda water. The example shown here was filled in the 1890s, but earlier bottles must have looked much the same when sold. It is likely that the soda water still contained in this bottle has lost its effervescence because the iron wire which held the cork has rusted away and allowed the cork to loosen. Additionally the cork has dried out because the bottle was stored for many years in a wooden crate which held it in an upright position.

A novel use for Hamilton bottles was found by the British Army serving in India in the mid-nineteenth century. A temporary shortage of regulation water-bottles obliged the soldiers to use Hamilton bottles as replacements. Writing home in 1857 an infantryman reported:

'We have been issued with common soda water bottles in leather cases and with shoulder straps.'

Many seem to have preferred the glass Hamilton bottle to the conventional metal water bottle then in use; they became standard issue in several regiments. Examples can be seen in the National Army Museum and in the Somerset County Museum, Taunton.

Plate 48 – It was Hiram Codd's globe-stoppered bottle with its unusual neck crimpings which took most of the Hamilton bottle market after 1872. This specimen was made between 1876 and 1881, the years during which Codd was junior partner to Ben Rylands at the Stairfoot Works at Barnsley. The bottle is embossed CODD'S PATENT. LONDON. RYLANDS & CODD. MAKERS. STAIR-

FOOT,BARNSLEY. Similar bottles made between 1881 and 1884 when Ben Rylands' son, Dan, was Codd's partner are embossed CODD & RYLANDS. Both types are rare.

The use of a spherical stopper inside the bottle had already been tried with limited success by other inventors before Hiram Codd used the idea in his patent of 1872. The secret of the Codd's bottle's success was the additional use of an india-rubber washer which fitted into a 'safety groove' just inside the lip. Gas released from the effervescent contents pressed the marble hard against this washer and thereby created a leak-proof seal.

The bottle's disadvantages were that the marble in the neck attracted the attention of small boys who broke countless numbers of bottles to obtain free marbles, and that the bottle was very difficult to clean because the restricted neck obstructed brushes used to clean the inside of the bottle before re-filling. Despite these disadvantages the bottle became the world's best-selling mineral water bottle.

Plate 49 – Some bottlemakers who made Codd's bottles under licence used part of a Hamilton bottle mould to make the body. Such bottles are called 'Codd-Hamilton hybrids' by collectors. Those which were made with flattened bottoms like the example shown here are called 'flat-foot Codd-Hamilton hybrids'. Both types are rare. It was of course quite unnecessary for a Codd bottle to have a rounded end because the bottle did not have to be stored on its side.

Plate 50 – Rival inventors competed with Hiram Codd and Dan Rylands throughout the late nineteenth century. This bottle, patented by J. Deeks in 1885, was very similar to Dan Rylands' 'Bulb' Codd, but the Deeks' bottle had only one marble recess in its shoulder.

Plate 51 – This bottle, a Haines' patent of the early twentieth century, had an elongated marble. It was one of the last internally-stoppered bottles used by British mineral water companies. Ironically it was advertised by its makers as 'the bottle of the future'. Although its oval marble was a deterrent to small boys, the bottle retained the other great disadvantage of internally-stoppered bottles – it was far too unhygienic for twentieth century customers.

Plate 52 – One of Dan Rylands' coloured-lip 'anti-theft' Codds with a red lip. Although the bottles proved popular with mineral water makers in the British colonies they were not a practical solution to the problem of bottle thefts in Britain's large towns and cities. Many examples now in collections have come from dumps in Australia, South Africa, and New Zealand.

Plate 53 – A blue lip applied to a Rylands' 'Bulb' Codd. Mineral water makers could specify blue, red, amber, or dark green lips when ordering and the lips could be applied to any of Rylands' globe-stoppered bottles. It was even possible to order coloured lips on bottles of a different body colour, but no examples have yet been found by British dump diggers. However, they have turned up in colonial dumps in Australia where the rarest finds have been blue-bodied Codd-Hamilton hybrids with dark green lips. Such bottles would have cost the mineral water makers who used them less than sixpence each to buy; a richly embossed specimen sold in 1975 fetched £200.

Plate 54 – Rylands' rivals competed by producing Codds in different colours. This dark green specimen from a dump near Birmingham, England, is an 'Expired Patent Codd Original', which means it was made at least fourteen years after Hiram Codd patented the 'Original' in 1872. Once patent protection expired any bottlemaker could produce the bottle without paying a licence fee to the inventor. In fact, Codd granted free licences to all manufacturers after the break-up of his partnership with Dan Rylands in 1884, subject only to the condition that they purchased the rubber rings used in the necks of the bottles from Hiram Codd's company.

Plate 55 – Hamilton bottle makers also offered their few remaining customers coloured bottles in the 1880s. Rarest are cobalt blue specimens like the attractively embossed example from Newcastle-on-Tyne shown here. The trademark is three fishes within a circle. It has been noted by dump excavators in Britain that richly coloured mineral water bottles are more frequently found in the dumps of industrialized cities and towns, especially in the north-east and north-west of England, and that it was in these regions that Hamilton bottles remained popular in the face of competition from internally-stoppered bottles.

Plate 56 – An olive-green Hamilton used by the Globe Mineral Water Company, Salford, Lancs. The company used a number of different mineral water bottles including Codds and 'bullet' stoppered varieties, all of them made in olive-green glass and all richly embossed with the company's ornate trademark.

Plate 57 – In the 1890s the 'improved' Hamilton, which was stored on its side but which could stand upright on a table, was introduced. Most of these bottles were made in aqua glass but a few, like this rare blue specimen, were produced in other colours. Although they were not widely used outside Britain substantial numbers of empty specimens appear to have found their way to North America in the early twentieth century as ships' ballast. Many embossed with the trademarks of Belfast mineral water makers have been recovered by skin-divers from rivers and harbours in Canada and the United States.

Plate 58 – The use of coloured marbles in Codd bottles achieved some popularity in the 1890s. The three aqua Codds shown here have amber marbles. Black, red, blue, dark green, and brown marbles were also used. It is not known whether they were employed as a method of identifying bottles or simply to make the bottles look more attractive to customers. No doubt they were also highly attractive to small boys hunting colourful marbles, a fact which must have deterred many mineral water makers from using them. It is certainly the case that nineteenth century refuse dumps containing coloured marble specimens also hold a very high proportion of broken Codds.

Plate 59 – These two 'bullet-stoppered' bottles had internal stoppers which were not held by strictures in the neck. At least twenty different bullet-stoppered bottles were patented and many, like these olive-green specimens, were made in coloured glass. Like Codd bottles they were overtaken in the early years of the twentieth century by screw-stoppered bottles. Although they were not prey to small boys hunting marbles, and even though their insides could be washed without difficulty their use was frowned upon by public health authorities because the stopper, which was exposed to dirt and germs when in the 'closed' position, fell into the contents when the bottle was opened.

Plate 60 – Internal screw-stoppered bottles eventually captured the entire mineral water bottle market and held it in Britain until very recently, though modern screw-stoppered bottles were made in aqua glass and rarely embossed.

Plate 61 – An extremely rare early twentieth century screw-topped bottle with a valve in its ebonite stopper. Before opening the bottle the customer pressed the valve to release the gas pressure, thus preventing loss of the contents by frothing. Very few of these unusual bottles were made, probably because the expensive stoppers were far too frequently lost by customers who failed to replace them before returning their empties.

Plate 62 – 'Lightning' or 'swing-stoppered' bottles were not widely used in Britain, but a few mineral water makers did employ them in the 1920s. Their great defect was that the wire bales soon rusted and made the bottles unsightly.

Plate 63 – All these bottles were used in the nineteenth century by a single British mineral water maker, A. Craven Ltd., of Hulme, Cheshire. The company's trademark, a seated figure of Britannia, is embossed on all the glass bottles and transfer-printed on the stoneware ginger beer. Included within the trademark are the words ESTABLISHED 1842. It is likely the company also used incised ginger beer bottles before 1850. Shown in this group are a Codd, a Hamilton, a flat-foot Hamilton, a 'bullet-stoppered' bottle, a dark green beer, an aqua beer, and a stoneware ginger beer. Many collectors try to obtain complete sets of bottles used by one company, especially when the bottles carry attractive trademarks as do those in this group.

Plate 64 – The only internally-stoppered mineral water bottle to achieve widespread popularity in the United States in the nineteenth century was the 'Hutchinson' bottle which had a rubber washer trapped in its neck which was pulled into a closed position after the bottle was filled by means of a wire loop protruding from the mouth. Shown here are a clear glass and a blue-green specimen.

Plate 65 – Two early chemist's shop display bottles. The clear flint glass specimen on the left is pontil scarred and has a ground stopper.

The dark green specimen is also pontil scarred and has a pewter-topped cork stopper. Both bottles are of approximately one pint capacity. They came from the cellars of an old chemist's shop which had been in business on the same site for more than one hundred and fifty years. Along with large quantities of out-of-date stock they had been dumped in the cellars and covered over the years by a mountain of empty boxes and cartons. The fortunate collector who gained permission to search the cellars and to keep whatever he found was assured by the present proprietor that there was 'nothing worth finding down there'.

Plate 66 – A group of early Victorian medicine bottles from Britain. *Back Row; Left to Right:* 1. Dalby's Carminative vial with pontil scar. 2. Ruspini's Styptic in blue glass. 3. Another Dalby's Carminative bottle with original contents, label, and waxed stopper. 4. Henry's Calcinated Magnesia bottle in clear flint glass with crude pontil scar. *Front Row; Left to Right:* 1. Robert Turlington's Balsam of Life. This bottle was one of several copied by Dr. Dyott at his American glassworks in the early nineteenth century. His counterfeits are thought to have been made in pale olive glass. The lips of Dyott's bottles were usually thinner than the lip on this British-made specimen. 3. The crude pontil scar on the base of this bottle indicates a date of manufacture before 1850. The word LONDON is embossed on one panel.

These bottles came from the same cellars which held the specimens seen in Plate 65. Such locations, together with the cellars and outhouses of nineteenth century houses, provide rich hunting grounds for bottle collectors who live in large cities and towns where there are probably as many unburied antique bottles still awaiting discovery as there are in unexcavated refuse dumps.

Plate 67 – One of the most widely counterfeited of all patent medicines, Daffy's Elixir was copied by both British and American medicine vendors during its many years of popularity. Most of them managed to obtain bottles that fairly closely resembled the bottles used by the rightful owners of the brand name, but there were several variations. In the 1840s the rightful owners were Dicey & Co. (late Dicey & Sutton) of Bow Church Yard, London. The company's name and address was embossed on the bottle, together with the words SEE THAT THE NAME DICEY & CO IS IN THE GOVT. STAMP OVER THE CORK. By the 1860s the words had changed to UNLESS THE NAME OF DICEY & CO IS IN THE STAMP OVER THE CORK THE MEDICINE IS COUNTERFEIT. Additionally the number 10 had been added to the street address. One of these later bottles is shown on the left of the photograph. The bottle on the right, a light olive specimen, has in addition to the address, the words SEE THAT THE WORDS DICEY & CO ARE PRINTED IN THE STAMP. The bottle has a pontil scar and is thought to have been made by Dr. Dyott, the American patent medicine and bottle maker known to have counterfeited the medicine. Another counterfeit in dark brown glass is shown in the centre of the photograph. It is embossed only with the words TRUE DAFFY'S ELIXIR. The cylindrical bottle in the foreground is also thought to be a counterfeit even though it is embossed with the words found in the genuine bottles. Other British 'pirates' who sold Daffy's Elixir and embossed their own names on the bottles include James Barclay (late Jackson & Co.), and J. Staples, both London wholesale chemists. There were probably others whose bottles have yet to be found.

Plate 68 – An unusual three-sided patent medicine from Britain. The bulb at the neck is thought ot have been a pouring aid. The bottle is embossed FISHER'S SEAWEED EXTRACT.

Plate 69 – Another British bottle of unusual wedge-shape. It is embossed PRICE'S PATENT CANDLE COMPANY LIMITED and bears a registered design mark. It once held glycerine-based cough medicine – (Glycerine was a by-product in candle making) – which sold widely in Britain and the colonies. The bottles, in both blue and aqua glass, have been found in Britain, Australia, New Zealand, and South Africa.

Bottles bearing diamond-shaped registered design marks are highly prized by collectors, not only because they are almost always of unusual shape, but also because the coded symbols incorporated in the mark enable the bottle's earliest possible year of manufacture to be ascertained. The diamond mark was used from 1842 until 1883. For designs registered between 1842 and 1867 a letter of the alphabet will always be found at the *top* of the diamond; for designs registered between 1868 and 1883 a letter of the alphabet will always be found at the *right hand side* of the diamond. The following list of code letters and dates should enable the reader to place an earliest possible year of manufacture on any bottle bearing a diamond registration mark.

1842–1867. (Code letter at *top* of diamond.)

EARLY

A – 1845	J – 1854	S – 1849
B – 1858	K – 1857	T – 1867
C – 1844	L – 1856	U – 1848
D – 1852	M – 1859	V – 1850
E – 1855	N – 1864	W – 1865
F – 1847	O – 1862	X – 1842
G – 1863	P – 1851	Y – 1853
H – 1843	Q – 1866	Z – 1860
I – 1846	R – 1861	

1868–1883. (Code letter at *right hand side* of diamond.)

A – 1871	L – 1882
C – 1870	P – 1877
D – 1878	S – 1875
E – 1881	U – 1874
F – 1873	V – 1876
H – 1869	W – 1878
I – 1872	X – 1868
J – 1880	Y – 1879
K – 1883	

Note: From 1884 the diamond mark was abandoned. Instead progressive numbers preceded by the words: 'Registered Design Number' were employed. No.1 was registered in January 1884; No.351202 in January 1900.

Plate 70 – Two late nineteenth century American patent medicine bottles. The dark green specimen held 'Dr. Townsend's Sarsaparilla', a famous 'blood purifier'; the amber specimen held 'Fenner's Kidney and Backache Cure'.

Plate 71 – The bottles used to package the medicines of H. H. Warner, one of America's most successful nineteenth century quacks, are familiar to diggers on both sides of the Atlantic and in Australia. Warner had been a salesman for a safe manufacturer before he turned to making patent medicines and he used a safe as his trademark. It is embossed on all of his 'Safe Cure' bottles. His New York business attracted so much custom he was obliged to open offices in London and Melbourne to cope with overseas orders. The bottles used to package the medicines sold overseas were made during the twenty-odd years in which the public consumed the product, in a number of British glassworks. These were additionally embossed with the word 'London' or 'Melbourne'. (A few specimens embossed 'Frankfurt' have also been found but it is not known whether these were British-made.) Because the bottles were made by various manufacturers they display slight differences in shape and colour. Shown here are six specimens from London, three brown and three green, made between 1880 and 1900. The largest green specimen is of two-pint capacity, but most of Warner's bottles were made in pint and half-pint sizes. Note the brown bottle on the extreme right; it contained 'Warner's Safe Nervine'.

Plate 72 – An extremely rare miniature green Warner's bottle shown here alongside a half-pint brown 'Nervine'. The few miniatures found in Britain have all turned up in early twentieth century dumps and it has been noted by researchers that at that time the London company advertised 'Warner's Safe Compound'. It is possible this was a concentrated form of the earlier medicines and therefore sold in smaller bottles which continued to be embossed with the word 'Cure'.

Plate 73 – William Radam's 'Microbe Killer' which, according to the embossing on its bottle, 'cured all diseases', was another American patent medicine consumed by an international market. The trademark shows a skeleton being attacked by a man wielding a club. In addition to the ornate embossing Radam's bottles were also labelled as shown. The label included exaggerated claims about the miraculous cures worked on those who swallowed the contents of the bottle; but when the medicine was examined by analysts working for the U.S. Government it was found to consist largely of water and acid.

Plate 74 – The manufacture and sale of bitters was an important branch of the American quack medicine market in the latter half of the nineteenth century. The 'medicine' was little more than alcoholic spirit flavoured with a few bitter-tasting herbs which gave it its name. Consumption of bitters was greatly encouraged by 'anti-liquor' campaigners who blamed whisky and other hard drinks for most of the social evils of the times. Bitters were exempted from criticism because those who consumed them could claim they were curing their aches and pains. In truth they were often drinking even stronger hard liquor than those who patronized saloons and bars.

Shown here are three figural bitters bottles: 'National Bitters', which was sold in a bottle shaped like an ear of a corn; 'Drake's Plantation Bitters' which was sold in a 'log cabin' bottle; and 'Indian Queen Bitters' which was packaged in a figural bottle in the shape of an Indian maiden.

Plate 75 – The manufacture and sale of hair restorer provided another lucrative market for nineteenth century medicine vendors. Thousands of brands were available, all claiming to promote the growth of new hair on bald heads, and most of them consisted of nothing more than oil, alcohol, and cheap perfume. Shown here are three typical American specimens in brown, aqua, and blue glass. The 'Mexican Hair Renewer' on the right was widely used; the bottles have been found in British, Australian, South African, and North American dumps.

Plate 76 – The sale of babies' feeding bottles was an important part of the chemists' sundries market and a wide variety of types and makes were sold. Shown here are two rare specimens: a boat-shaped earthenware feeder of the early eighteenth century. Milk was poured into the open top and the bottle was used without a teat. Some of these early earthenware types were decorated with blue patterns. The purple specimen bears the date 1871 on its base. It is thought to be of American manufacture and originally made in clear glass purpled by the action of the sun. The screw cap is pewter with a central tube holder made in ivory. A piece of rubber tubing passed through the hole into the bottle. A teat was fixed to the other end of the tube and this was placed in the baby's mouth. Children were not usually nursed during feeding in the nineteenth century; the bottle was placed on a stool or table near the baby's cot and the child sucked its milk through the tube.

Plate 77 – A group of British baby feeders. The upright model at the rear is embossed MATHER'S INFANT'S FEEDING BOTTLE and has a ceramic screw top and a glass tube reaching into the body of the bottle. The elongated and single-ended model on the left is embossed PICCANINIES O.F. FEEDER. The letters O.F. are probably an abbreviation of 'old fashioned' because this type of bottle went out of general use in the 1860s. The model on the right has an anchor trademark embossed and uses a glass internal screw stopper. This type was most widely used in Britain between 1880 and 1900. The 'banana' feeder, a double-ended model, came into widespread use early in the twentieth century.

Plate 78 – The use of blue glass as a means of identifying poison bottles became widespread in the late nineteenth century, though not all bottles for noxious substances were blue. Shown here are a group of poison bottles used by British chemists between 1880 and 1910. *Back row; left to right:* 1. Six-sided blue poison embossed NOT TO BE TAKEN and with deep vertical ribbing on unembossed panels. This was the most common poison bottle used in Britain; it was also made in dark green and (rarely) dark brown glass. 2. A three-sided blue poison embossed NOT TO BE TAKEN and with raised projections embossed. This shape was usually reserved for small poison bottles of less than 4 oz capacity. 3. A pale blue poison bottle embossed POISON. NOT TO BE TAKEN and with a deep indentation in its body. The design was patented in 1905. 4. An eight-sided poison bottle embossed NOT TO BE TAKEN and with indentations between the panels and embossed ribbing. *Front row; left to right:* 1 and 2. 'Submarine' poison bottles. They enjoyed a brief period of popularity in Britain at the turn-of-the-century and were designed to be identified by their unusual shape. Green specimens were also made. 3. A 'Martin' poison bottle with a deep indentation below the neck which limited the dosage dispensed when the bottle was tilted. This bottle was also easily identified by touch.

Plate 79 – A large British poison bottle of 20 oz capacity. Made in 1910.

Plate 80 – American poison bottles were also often made in cobalt blue glass in the nineteenth century. Shown here are two figural poisons; one in the shape of a skull, the other in the shape of a coffin. Both were widely used in the United States before 1900.

Plate 81 – Attractive turn-of-the-century perfume from Britain. The case pair in the centre were issued by the Vinolia Company of London in the 1890s. The Crown Perfumery bottle fourth from left has a crown-shaped stopper; this was registered as the company's trademark and stoppers of this shape were used on all of its bottles. The bottle second from left held 'Larbelestier's Original Jersey Queen Eau de Cologne as supplied to the late Queen Victoria'.

Plate 82 – Opaque milk glass bottles were widely used in the nineteenth century for cosmetics and toiletries. Shown here are three pewter-topped green specimens marked MOUTH WASH, LOTION, and ASTRINGENT, together with two white bottles used for hair preparations.

Plate 83 – An embossed blue milk glass bottle from an Australian dump.

Plate 84 – Both men and women carried pocket spitoons in the nineteenth century and 'spitting flasks' were widely advertised as druggists' sundries. The blue glass specimens shown here all have metal fittings of pewter. The mug in the centre was probably a 'bedside' version.

Plate 85 – A selection of bottles including poisons, colognes, and perfumes once sold in chemists' shops throughout Europe. They resemble British and American bottles used for the same purposes, but Continental bottlemakers seem to have used a wider range of colours. In addition to embossing and paper labels two other methods of marking pharmaceutical and cosmetics bottles were widely used:

acid-etching, as seen on the silver nitrate (Ag NO3) bottle, and enamelling as seen on the Sirop de baume du toilu and Essence d'origan bottles. The acid-etching was carried out with hydrofluoric acid applied to a stencil held against the body of the bottle. Chemists often applied names and chemical symbols to plain bottles in this way; the resulting 'label' was both water resistant and impervious to stains. Enamelled labels, often applied to bottles likely to be used in bathrooms where steam would have caused paper labels to fall off, were reserved for expensive products and special display bottles used in chemists' shops. Lettering was either hand-painted on the surface of the glass or painted in reverse on thin glass panels which were adhered to the bottles after the enamel had hardened.

Plate 86 – Not the interior of a nineteenth century shop, but part of the superbly displayed collection owned by Ben Swanson, an American enthusiast at present living in Britain, who has specialised in containers used by Victorian chemists. In addition to an incredible array of medicine bottles (many with their original contents and labels) the shelves are stacked with chemists' sundries including perfumes, toiletries, bed-pans, inhalers, food warmers, medical appliances, and what is probably the world's largest privately owned collection of transfer-painted toothpaste pot lids. (For further information on pot lids see my earlier book, *Collecting Pot Lids*, Pitman, 1974.)

Plate 87 – Another view of Ben Swanson's collection showing many examples of enamelled and acid-etched labels used on chemists' shop bottles.

Plate 88 – Five rare inks from Britain. *Left to right:* 1. A domed specimen with offset neck embossed BLACKWOOD'S PATENT, LONDON. The sloping neck enabled the user to reach the last few drops of ink without tilting the bottle. Pale blue examples of this bottle have also been found. 2. A figural cottage ink in aqua glass. The protruding end window is unusual; most British cottage inks have flat windows. Both blue glass and brown stoneware specimens have been found in Britain but they are extremely rare. Most have turned up in dumps dated earlier than 1880. 3. A brown stoneware teakettle ink incised MORREL LONDON. The rear of the bottle carries a registered design number. A few glass teakettle inks have been found in Britain, but the shape is much commoner in the United States. As with domed specimens the sloping neck made it easy to reach the last drop of ink in the bottle. 4. Another type of figural ink fairly common in America but rare in Britain – a barrel with embossed hoops. 5. A triangular ink complete with original contents and stopper; embossed DERBY ALL BRITISH. Blue and brown glass specimens have also been found.

Note that all four glass inks have sheared lips, a feature of almost all British glass inks made in the nineteenth century. The effectiveness of this type of closure when used with a cork and a little sealing wax is seen in the triangular specimen which has remained leak-proof for more than a century.

Plate 89 – Three attractive figural inks from the United States. The stoneware specimens were issued by the Carter Ink Company and are known to collectors as 'Ma and Pa Carter inks'. They were sold between 1900 and 1930 and pairs are highly prized by present-day collectors. The glass specimen was made in the late nineteenth century and is thought to represent the head of Benjamin Franklin.

Plate 90 – An unusual metal inkstand holding a pair of figural snail-shell ink bottles. Thought to be of American manufacture, the stand is marked with the patent date 25 November 1879. Figural inks in the shape of snail shells were sold in America up to the 1920s.

Plate 91 – A bulk ink of half-pint capacity made for the Stafford Ink Company of America in the 1890s. Although made in glass the shape of the bottle is very similar to the stoneware bulk inks in which most European ink makers exported their products to America throughout the nineteenth century. The designer probably kept to that shape because that was how the public expected bulk ink bottles to look; but the eye-catching blue glass must have won many customers away from salt-glazed stoneware.

Plate 92 – An interesting bulk ink bottle with original contents, stopper and label. The label was used in 1900 by a Greek ink manufacturer, Menounos of Athens. It is printed in both French and Greek and also carries the words TRADE MARK in English. English words are more likely to have been used for the benefit of customers in America where bulk ink bottles made in glass were commonly found; British ink makers used stoneware exclusively for large capacity ink bottles.

Plate 93 – French ink bottle of the 1890s in aqua glass; embossed J.GARDOT,DIJON. Although not so common as eight-sided, square, and oblong inks, this shape was also widely used in Britain at about that time. Note the neatly finished lip which suggests the bottle was machine-made. Similar lips are not found on British inks before 1910, though they are often seen on American-made bottles of the late nineteenth century.

Plate 94 – Glass fairy lights once decorated most Victorian Christmas trees. They were sold in sets of several different colours and before hanging on the tree with wire handles tied beneath the ring below the lip a lighted candle was placed inside each one. The effect was a delicate glow of colour which must have delighted many a Victorian child. The specimens shown here are of the commonest type with 'basket weave' embossing. Other designs, including rare specimens in the shape of Queen Victoria's head, have been found.

When fairy lights were displaced by electric lights for Christmas trees at the beginning of the twentieth century thousands of complete sets were thrown into dustbins. It is for this reason that dump excavators who find one specimen usually turn up several more in other colours within a few yards of the first find. Complete sets are occasionally sold by antique dealers.

The most prolific maker of fairy lights was the firm of Pattison & Co. of London whose name is often found embossed on bases. The same company also made many glass oil lamps.

Plate 95 – Oil lamps from British and Australian dumps. A brass collar incorporating a wick holder was fitted around the crudely finished neck; the wick hung down into the body of the lamp which was filled with oil. The introduction (in England) of gas lighting for homes in the late nineteenth century killed the market for oil lamps in cities and towns, but they were used in country districts until the 1920s. Cheaply manufactured in two-piece moulds, they sold, complete with brass collar and wick, at sixpence each in London in 1880. Restored specimens with glass chimneys added are often seen in antique shops but digging evidence suggests that they were used in the nineteenth century without chimneys which are *not* found in dumps. Original brass collars do not survive long burial.

The light blue specimen in the centre of this group is uncommon; most oil lamps were made in aqua glass.

Plate 96 – Another use for glass containers which died out in the early part of the twentieth century was as fire extinguishers used in hotels and offices. Filled with carbon tetrachloride, the extinguishers were mounted on hotel and office walls in wire holders. In the event of a fire they were taken from the holders and thrown into the centre of the flames where they shattered and smothered the fire with their chemical contents. The fumes given off by carbon tetrachloride can be lethal if inhaled in large doses and this is one reason why glass 'fire grenades' were superseded in the 1920s by hoses and buckets of sand. Shown here are two MINIMAX grenades in bright green glass. Blue and red grenades, some of spherical shape, were also made.

Plate 97 – Shown in these two photographs are three unusual figural bottles from Poland. The uniformed figure is thought to be a Polish coal miner in a traditional costume. The bottle is pontil scarred and dates from the mid-nineteenth century. The pipe-shaped bottle has a glass and cork stopper which forms the pipe's mouthpiece. The slipper, another pontil scarred specimen, also appears to have used a cork stopper. All three bottles are thought to have held Polish vodka.

Plate 98 – A delightful Spanish figural bottle moulded in the shape of a church. It is pontil marked and thought to date from the 1860s. Original contents unknown, but a collector in Spain reports that in the nineteenth century holy water was sold at sacred shrines in bottles in the shape of crucifixes and other religious devices. Perhaps this church figural had a similar origin.

Plate 99 – A superb bear figural in what at first appears to be black glass but which, when held up to strong light, exhibits a deep purple colour. It is thought to have been made in Germany in the nineteenth century. The bear has a shield around its neck to which a label could have been applied. It is possible this bottle was originally made in clear glass to which a large quantity of manganese was added and that later exposure to sunlight turned the glass purple. Sun-coloured bottles in shades of purple as deep as this are not unknown in the desert regions of America and Australia. Alternatively the glass might have been coloured at the glassworks by the addition of an oxide of uranium. This method of producing purple glass was known in the nineteenth century, though it was usually reserved for the manufacture of expensive wine glasses.

Plate 100 – A figural liqueur bottle in the shape of a swan made in Greece in the 1920s when the Art Deco style was at the height of its popularity in Europe. Impractical design was a feature of many Art Deco items and this bottle, which is over twelve inches tall, certainly qualifies on that score. The neck of the swan is solid and must have served as a handle when pouring the contents. Swan shaped bottles have been used more recently in the United States and Britain as containers for perfume, but they have all been of very small size.

Plate 101 – Another example of Art Deco style; a bow-shaped British perfume bottle in cobalt blue glass dating from the 1920s.

Plate 102 – The collecting of twentieth century bottles is a popular branch of the hobby in the United States, partly because many of the country's nineteenth century refuse dumps have been totally excavated by diggers, and partly because the American packaging industry produces an abundance of eye-catching bottles. Shown in this group of photographs are perfume bottles, miniature spirit bottles, hair dressing bottles, and confectionery bottles – all made quite recently and all enthusiastically sought by present-day collectors.

Plate 103 – A colourful international display showing examples from Britain, America, Europe, and Australia; all once discarded as worthless empties; all recovered by dump diggers to be carefully cleaned and proudly displayed on collectors' shelves. Specimens shown here fetched a total of £300 in an antique bottle auction held in Britain in 1975. *Left to right:* Dark green Codd (British); Hair restorer (American); Cough mixture (British); Liqueur (Dalmatian); Quack medicine (Australian); Sealed wine (British); Schnapps (Dutch); Cobalt mineral water (British).

Forming a collection of antique bottles

Bottle collecting is one of those very few branches of antiques at which the beginner can make a start without spending much money. Indeed, anyone who possesses the strength to dig and the determination to find a suitable site can fill his display shelves with prized specimens for little more than the cost of a few handtools. Even the non-digging enthusiast who buys his bottles can pick up wonderful bargains if he shops in the right places and concentrates his attention on specimens in the lower and middle price ranges.

After reading the books and magazines mentioned in the following lists of recommended reading, the next step should be to join a bottle collectors' club. In Britain there is a national organization, The British Bottle Collectors' Club, which has branches in every county. The small membership fee entitles each member to attend local monthly meetings where auctions, exchanges, talks on bottle history, and the sale at very modest prices of bottles found by local diggers are regular features. For newcomers interested in site excavations there are opportunities to learn the techniques at first hand by joining in local group digs organized by the Area Secretary. Each digger keeps the bottles he recovers, though members are expected to pass on details of the bottles they dig up to the Area Secretary so that a comprehensive record of finds is maintained. The only tools required on these excavations are a garden fork and a shovel, and the cost of reaching the site is shared by all participants. Thus can any newcomer make a very inexpensive start to the hobby in Britain.

In America and Australia each state has its own club. Meetings of members are less frequent than in Britain, but when such an event takes place it is usually an extravagant three-day 'show' which includes competitions for the best displays, auctions of high-value bottles, and a sale at which bottle lovers from all parts of the country hire stalls and sell thousands of specimens. These shows are widely advertised months in advance in nationally circulated bottle magazines to ensure large attendances. Small events organized on similar lines take place in Canada, South Africa, and New Zealand.

In Britain the non-digging collector seeking uncommon bottles might visit any one of a dozen large shops devoted exclusively to selling bottles. Prices in these establishments are somewhat higher than one would expect to pay if buying bottles at a local club meeting; but there is no doubt the rarest and most highly prized specimens are usually to be found there. Another method of obtaining rarities is to scan the sales columns of magazines such as *Old Bottles and Treasure Hunting*, a market place used by many diggers who live in remote regions and who are therefore unable to attend monthly club meetings. As in other branches of antique collecting, bargaining is freely practised; most sellers ask rather more than they expect to receive for a bottle and are prepared to haggle before making a sale. There is less scope for price adjustments in the United States where the publication of numerous price catalogues has firmly established the selling prices of many rare bottles. Nevertheless, the collector who owns duplicates of rarities usually has little difficulty in finding a seller prepared to consider an exchange, and bottles are frequently exchanged internationally by collectors in America, Britain, Australia, and South Africa.

Because the hobby is relatively new there is vast scope for the beginner prepared to concentrate on inexpensive bottles, many of which are greatly under-priced because the majority of collectors have specialized in sealed bottles, historical flasks, coloured mineral water bottles, and other rarities. There can be no doubt that in five or ten years, when most of the easily accessible sites have been excavated, the values of these 'common' bottles will rise sharply. Now is the time to build a comprehensive collection.

Book and magazine list

BOOKS:

A Treasure Hunter's Guide, Edward Fletcher, Blandford Press, Poole, U.K.

American Glass, G. McKearing, Crown Publishers Inc., New York, U.S.A.

Bottle Collecting, Edward Fletcher, Blandford Press, Poole, U.K.

Bottles in Australian Collections, James Lerk, Cambridge Press, Victoria, Australia.

Collectors' Guide to Seals, Case Gins, and Bitters, Edward Fletcher, Latimer, London, U.K.

Digging Up Antiques, Edward Fletcher, Pitman, London, U.K.

International Bottle Collectors Guide, Edward Fletcher, Blandford Press, Poole, U.K.

The Non-Dating Price Guide To Bottles, Pipes, and Dolls' Heads, Edward Fletcher, Blandford Press, Poole, U.K.

Treasure Hunting For All, Edward Fletcher, Blandford Press, Poole, U.K.

MAGAZINES:

Australian Bottle Review, P.O. Box 245, Deniliquin, N.S.W., 2710, Australia.

International Bottle Trader's Gazette, 104 Harwal Road, Redcar, Cleveland, U.K.

Old Bottles and Treasure Hunting, 801 Burton Road, Midway, Burton-on-Trent, Staffs, U.K.

Old Bottle Magazine, Box 243, Bend, Oregon, 97701, U.S.A.